Values in Landscape Architecture and Environmental Design

READING THE AMERICAN LANDSCAPE

Lake Douglas, Series Editor

"Borrowed landscape" in the arboretum of the University of Illinois, Urbana-Champaign, seen from the garden of Japan House. Photograph by M. E. Deming.

Values in Landscape Architecture

FINDING CENTER IN THEORY AND PRACTICE

and Environmental Design

Edited by **M. ELEN DEMING**

LOUISIANA STATE UNIVERSITY PRESS

BATON ROUGE

Publication of this book is made possible in part by support from the Brent and Jean Wadsworth Fund of the Department of Landscape Architecture, University of Illinois, Urbana-Champaign.

Designer: Barbara Neely Bourgoyne
Typefaces: MillerText, text; Cinta, display
Printer and binder: Thomson-Shore, Inc.

A portion of "Beyond 'Sustaining Beauty': Musings on a Manifesto," by Elizabeth K. Meyer, was published previously in *Taylor Cullity Lethlean: Making Sense of Landscape* (Spacemaker Press, 2013), ed. Gini Lee and SueAnne Ware. Copyright © 2013 Taylor Cullity Lethlean.

"From Gotland to Youngstown: The Indissoluble Link between Landscape and Justice" is an abridged and modified version of "Landscape and Justice," by Tom Mels and Don Mitchell, in *The Wiley-Blackwell Companion to Cultural Geography* (Malden, MA: Wiley-Blackwell, 2013), ed. Nuala Johnson, Richard Schein, and Jamie Winders, 209–24. Copyright © 2013 John Wiley & Sons, Ltd.

"City of Nawabs to City of Elephants: Urban Transformation in Lucknow, India" was first published online, in different form, as "Mayawati and Memorial Parks in Lucknow, India: Landscapes of Empowerment," *Studies in the History of Gardens & Designed Landscapes* (July 2014): 1–16. An abridged version of this essay was also published in *Journal of Landscape Architecture* (India), 68 (2013): 68–76, and is used with permission of the editor.

Library of Congress Cataloging-in-Publication Data

Values in landscape architecture and environmental design : finding center in theory and practice / edited by M. Elen Deming.
  pages cm. — (Reading the American landscape)
  Includes index.
  ISBN 978-0-8071-6078-7 (pbk. : alk. paper) — ISBN 978-0-8071-6079-4 (pdf) — ISBN 978-0-8071-6080-0 (epub) — ISBN 978-0-8071-6081-7 (mobi) 1. Landscape architecture—Philosophy. 2. Landscape assessment. 3. Landscape ecology. I. Deming, M. Elen, 1956– editor.
  SB472.V285 2015
  712.01—dc23

                                   2015012002

CONTENTS

PREFACE AND ACKNOWLEDGMENTS

Since the inception of this project, concepts of "value" have surfaced and swirled in countless dizzying iterations for the team of authors and editors involved. The term equally connotes strategic or quantitative dimensions, as in "the value of investment in our institutions," as well as humanistic or qualitative dimensions, as in "the value of a college education." There is also the sense of plural or shared popular and/or political codes and stereotypes, as in "midwestern values," or "family values," which stand in for larger systems of behavior and belief. Because all synonyms for value imply worth, goodness, or virtue, they tend to invoke corollary shadings of ethics and morality and, especially when considered in relation to notions of center or norm, can generate a host of provocative ambiguities.

This volume of essays—an autonomous and yet intellectually porous collection— is the result of these variously explored interests. The purpose of this book is to establish several broad themes, along with specific cases and benchmarks for those themes, in order to provoke a concentrated, contemporary discourse on the value(s) of landscape—a topic that has been addressed episodically in the literature and yet has never been more important than it is today. In selecting this representative range of perspectives from among a wider variety of potential topics and voices, we hope that practitioners, students, and scholars in a variety of environmental design disciplines will find this volume expansive, evocative, and useful.

It has been humbling and encouraging to discover how many works on the subject of landscape and values—including both implicit and explicit treatments—have been undertaken since the early 1970s. The breadth and depth of that body of work is evident in the concepts and references developed within each of the following essays. While we certainly never imagined that this book would offer the last word on the subject, we feel fortunate and grateful that our thoughts will join that broader literature, which should and undoubtedly will continue to grow.

To possess any merit at all, volumes of this scope and complexity must be guided by many voices and shaped by many hands. This project is no exception. In recognition of their many contributions to this work since 2012, both tangible and intangible, warm thanks go to my colleagues in the Department of Landscape Architecture at the University of Illinois, Urbana-Champaign, especially David Kovacic, Stephen Sears, and Amita Sinha, all of whom were involved in the initial shaping of the theme. That theme was tested at a 2012 conference hosted by the University of Illinois for the Council of Educators in Landscape Architecture (CELA), and a separate call for papers was sent out afterwards, in summer 2012, to landscape scholars through many channels.

Contributing authors (comprising a wide variety of institutions and practices) are to be commended for the quality and significance of their commitment to this project, especially their kind endurance of the halting preliminaries, long editing and review process, and rigors of final production. Each of the essays has been substantially expanded, developed, and revised on the basis of two levels of editorial review as well as external

peer review. Some of the essays (Meyer, Moore, Sinha and Kant, and Mitchell and Mels) have been condensed, reframed, and/or updated from previously published material as acknowledged. An earlier version of the editor's introductory essay was shared at an international forum at Tsinghua University (October 2013), where audience members offered formative and constructive feedback. However, all of the authors have continued to thoughtfully advance their topics during the long gestation of the book, through their reflections on practice and teaching and in correspondence with others.

Partners in this volume are indebted to the LSU Press and its editors, especially Margaret Lovecraft and series editor Lake Douglas, who encouraged the project from the very beginning. LSU Press maintained a steady vision for the book and its potential audience, and provided an invaluable and constant reminder of why values matter. Although the series is entitled Reading the American Landscape, this collection expands that geographic scope considerably, encompassing Australia, India, Scandinavia, and the Netherlands, among other places. The press's editors also exercised admirable expertise, patience, and tact as our project met the inevitable bottlenecks of the academic calendar. The team of writers is collectively grateful to the anonymous external reader(s) who offered significant and critical feedback in just the right ratio of challenge and encouragement.

The editor recognizes the Brent and Jean Wadsworth Fund of the Department of Landscape Architecture (University of Illinois), which generously provided funds to underwrite this project.

Finally, the editor acknowledges her personal gratitude to several colleagues. Kathryn Moore provided exquisitely thoughtful drawings to illustrate nuanced points in the introductory essay, "Value Added." Elizabeth K. Meyer and Simon R. Swaffield reviewed early attempts at synthesis and suggested how it might be done better. All are longtime personal friends, mentors, and model scholars on the topic of values. Their fresh and thoughtful vantages on this topic were offered with considerable sensitivity and keen insight, and gave a helpful editorial push in the right direction when it mattered most.

Values in Landscape Architecture and Environmental Design

M. ELEN DEMING

Value Added

AN INTRODUCTION

Values in Landscape Architecture and Environmental Design: Finding Center in Theory and Practice offers an intimate look at the complex alchemy that renders values into places. Indeed, values matter precisely because they materialize. Values comprise a veiled matrix of attitudes and expectations transformed in a variety of ways into the ordinary and extraordinary landscapes within which we spend our lives. Because values motivate the development of social mechanisms such as property laws, environmental policy, and professional practices, as well as theories of design process and form, they serve to shape the world's landscapes, whether ordinary or iconic, here or there, now or later, in an ongoing and global process.[1]

In particular, values play a critical role in the design and management of landscapes, a point that James Corner makes explicit: "design can be seen as a value-laden activity that not only reflects, but also constitutes the ethos of a culture. We are what we build, and we appear to ourselves through our building" (cited in Rosenberg 1993, 49). In other words, both the ordinary and designed landscapes hide the values of their makers in plain sight. Through interpretation, aspects of landscape form can be transmuted back into the social values that guided their construction, thus illuminating the context of past and present societies and perhaps anticipating future landscapes. The objective of this book therefore is to awaken for readers a broad capacity for landscape literacy and to

1

suggest how that capacity might be exercised in their lives personally, professionally, and politically.

Landscape and Values

What is landscape? Landscape is a complex, even transcendent concept. Although subject to natural processes, landscape is neither to be confused with the concept of nature nor is it synonymous with residual wilderness. Most emphatically, landscape is *never* a verb referring to gardening or maintenance practices. Landscape is a noun—it is a geographical entity, place, or condition made by people.

For the past four decades or more, cultural geographers and landscape historians have advanced the argument that landscape is both an idea and a place—a physical environment (or its representation) that mediates the realms of nature and culture along an endless continuum of imaginative possibilities. Although landscape is multivalent and does not belong solely to either realm, what has become quite clear is that landscapes are constructed by and for humans. At its most basic level, landscape refers to a category of geographical space, but the human condition provides its *sine qua non*.

Physical landscape is humanity's most important commonwealth. Yet even as we unquestionably depend upon, measure, and evaluate its utility and performance, it remains poorly understood as a system of meaning. In much the same way that fish are probably oblivious to the concept of water, humans have difficulty comprehending landscape as a medium with which we have coevolved morally and imaginatively, as well as biologically. The result is that the vast majority of the world's population never considers how landscapes express social values.

In this book we build on the notion that landscape is both a material and a conceptual medium that expresses social beliefs of ownership, control, status, power, virtue, spectacle, beauty, and faith, among many other things. Some designed landscapes perform at high levels just as other great cultural works do—music, architecture, painting, poetry, and literature included. And similar to some forms of human sounds, shelters, markings, and utterances, not all landscapes are highly expressive. Yet no landscape is ever completely value-free. This point offers the clearest distinction between landscape and related categories of environment, and secures the closest tie between landscape design and all other forms of human expression.

Despite the wide variety of places and practices they discuss, all the essays collected for this book share certain major assumptions about landscape(s) and value(s).

The first assumption is that all landscape types—whether they are urban, suburban, agricultural, or residual—are shaped by both place-based and resource-based values. Further, because physical landscapes have reciprocal effects on the social formation of our personal and collective values, our landscapes also form *us*. As we shape a landscape, so too does it shape us in return, affecting our lived experience and thus continuously forming, affirming, consolidating, or changing our values.

Another assumption shared by these authors is that landscape values perform at multiple scales, even multiple locations, and simultaneously affect human, non-human, and even non-biotic interests alike. The air we breathe, food we eat, water we drink, places we inhabit, and parks in which we recreate are all heavily values-laden. Each is conditioned upon balancing ownership rights, public health and professional regulations, and environmental policies somewhere within the moral, imaginative, and pragmatic coordinates of society.

These essays focus principally on cultural landscape processes, especially on values that are time-based. From ideas rooted in the distant past, many enduring values bear heavily on contemporary landscape perceptions and beliefs and thus extend their long reach into the future. Interpretations of place- and time-based values may also work backward, as cultural landscape historians try to tease apart what once was valued from landscapes that were built—and what was hoped from what really happened. Landscapes evolve, with impacts or benefits that may not be felt for a generation or more.

Perhaps because human identity and existence are bound so closely to land, and partly because of the land's slow rhythms, it is easy to ignore most landscape as a background condition, an uninteresting, insignificant resource. Occasionally, and more problematically, society is reminded that landscapes are also subject to cyclical processes of nature. Although the predictable pulses of the environment may be inspiring and sustaining, its power surges and sudden cataclysmic adjustments can be devastating and heartbreaking. Having woefully short-term memories and even shorter-term imaginations, people are prone to forget the inevitability of these processes; however, wildfires in dry ranges, floods in river valleys, droughts in agricultural regions, hurricanes along coastal zones, and tsunamis along the shifting plates of the Pacific Rim periodically recur to teach those lessons anew. As a recipient of these pulses, landscapes are affected just as people are: they are repeatedly made and unmade by both natural and cultural processes.

Perhaps we forget to account for landscape processes precisely because landscape has such a powerful naturalizing effect on the value systems that shape it. When masked in this way, values embedded in landscape by human practices may be rendered invisible, *as if natural,* and are therefore easy to overlook. It takes endless practice to learn to see through landscape's deceptions. Sometimes humanistic methods can help us along. A familiar example of how ordinary landscapes embed social values is the iconic poem "Mending Wall" penned by the American poet Robert Frost (1874–1963):

> There where it is we do not need the wall:
> He is all pine and I am apple orchard.
> My apple trees will never get across
> And eat the cones under his pines, I tell him.
> He only says, "Good fences make good
> neighbors."
>
> (Frost 1964, 47)

Frost's poem discloses landscape values poetically rather than explicitly, but it speaks clearly of land ownership, responsibility, privacy, respect, shrewdness, neighborly collaboration, uncomplaining labor, craft, local resourcefulness, tradition. Reading the poem today gives a bittersweet recognition of the passing of those values along with that landscape and the men who made it. As a materialization of residual New England cultural values, these stone walls are now understated, gray, and covered with lichens.

The landscapes they once bounded are nearly invisible, having first been overgrown with second- and third-growth forests and later developed for suburban homes and the rest of the economic apparatus that serve them.

This is not to say that Frost's landscape is gone; rather, it has simply become blurred and unintelligible. To many residents of this landscape it appears to be nothing more than derelict post-agricultural wasteland. For others, however—the land developer, the forest historian, the child exploring the woods near a suburban home—such landscapes continue to give up long-held secrets even as they function as active manuscripts of new values, rewritten every day. "Mending Walls" demonstrates that landscape is a malleable yet enduring medium for expression, a synthetic artifact constructed at least in part by social actors or agents (ownership, industry, laws, design, neighborhood, and so forth) motivated by social values (control, productivity, utility, stewardship, beauty, and so forth). In short, nothing a society believes, imagines, does, or makes of the landscape ever escapes the influence of values—past, present, or future.

From this vantage, the function of the built environment as a repository of personal, regional, and even global values can be understood more clearly. Landscape agents who are values-aware can code or hide specific values in their work. Designers or

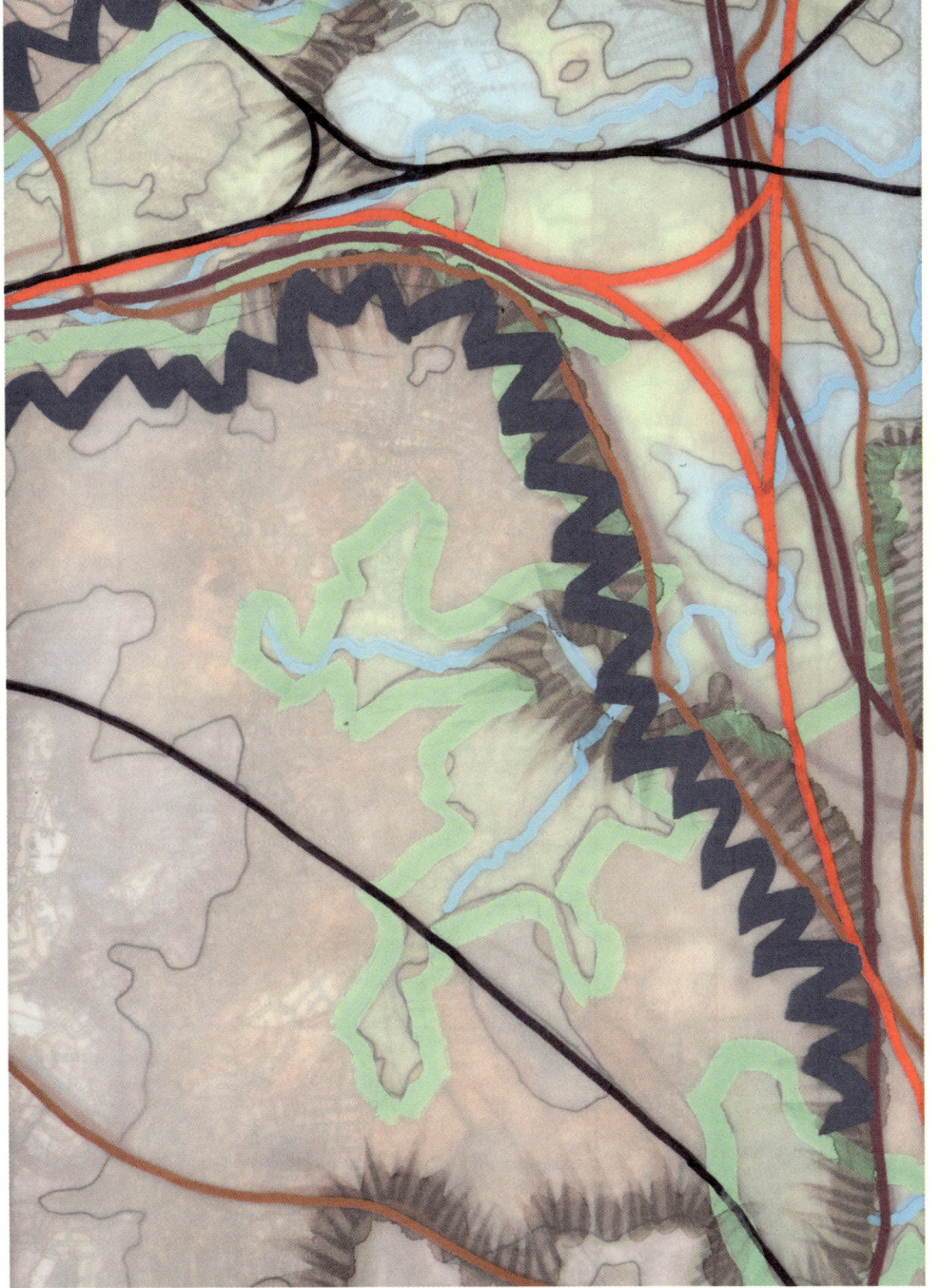

Fig. 1.1. Detail of diagram for the High-Speed 2 Rail Link Landscape Vision (HS2LV) master plan, showing values as selected and inscribed on landscape by planners and designers, 2012. Designed and drawn by Kathryn Moore. Clients: Birmingham City Council, Solihull MBC, Birmingham Chamber of Commerce, Centro, and Arups. Used by permission.

policymakers may "write" (inscribe) or "read" (decode and reveal) values in built, managed, and preserved landscapes (fig. 1.1). Informed consumers and citizens can and should critique both the writing and the reading, especially when putatively undertaken on their behalf. Any understanding of the regrettable history of ignorance, bias, and privilege that structures our landscapes should remind us that there is an effectively permanent imperative to be value-aware or values-literate. The cultural geographer Peirce Lewis sums it up: "Our human landscape is our unwitting autobiography, reflecting our tastes, our values, our aspirations, and even our fears, in tangible, visible form" (1979, 12). Whether we gain a level of landscape literacy or not, our values will inevitably continue to dictate the shape of our landscapes.

Landscape actors and designers are presented daily with opportunities to become aware of their own implicit values and acknowledge what or whose interests they may serve. To highlight these opportunities, the essays collected in this book present a wide variety of voices, positions, projects, geographic locations, and cultural perspectives on landscapes, landscape practices, and the values they express. The purpose of this collection is to magnify the social and environmental values hidden within a variety of landscapes, and to foster and strengthen values-literacy for readers. The book also provides a broad understanding of the social, philosophical, and environmental values guiding scholars, activists, and professional practitioners, particularly those working in the fields of landscape architecture, urban design, and environmental planning.

This rich tapestry of new and old cultural landscapes—from a modern city in India to a contemporary botanical garden in Australia, and from the leisure economy of Napa Valley, California, to the production of patriotic narrative in Washington, D.C.—demonstrates a range of methods for understanding social, political, environmental, temporal, and aesthetic values. All of these methods are highly transferable. Readers are therefore encouraged to apply methods and lessons from these studies to other projects, places, and populations they may be familiar with.

Concepts and Categories in Cultural Landscape Studies

This book contributes to the field of cultural landscape studies, broadly conceived, with framing concepts and terminology that are liberally drawn and synthesized from supporting disciplines. Although the limitations of this introduction do not afford an exhaustive review of cultural landscape scholarship, the list of works collectively cited by contributing authors very capably accomplishes that task. In particular, essays by Meyer, as well as Mitchell and Mels, provide a rich set of references. Readers new to landscape studies, however, may benefit from a brief orientation to a few now-classic works that have guided the discourse connecting landscape and values. Setting the table in this way also discloses the intellectual genealogy and literary timeline that this volume seeks to extend.

The best place to begin any tour of landscape studies is with the recognition and interpretation of socio-environmental values embedded in everyday, ordinary landscapes—in other words, landscape literacy. Among the early interpreters of cultural landscape is the midwestern naturalist May Theilgaard Watts (1893–1975). Even today, Watts's classic early work *Reading the Landscape of America* (1957; reprinted 1975) illustrates how landscape interpretation opens new understanding of values associated with landscape processes: "The land offers us good reading, outdoors, from a lively, unfinished manuscript. Records, prophesies, mysteries are inscribed there, and changes—always changes. Even as we read from some selected page, whether mountaintop, forest, furrow,

schoolyard, dune, bog, we see changes: in stirrings and silences, flavors and textures, spacing, tolerances, and confrontations and tensions at the edges" (Watts 1975, v).

Indeed, many people will make landscape interpretations in their everyday life, often without even questioning their thoughts. How many times has someone said, "That garden is lovely; the person who made it must be a nice person," or "what a dismal street; why don't the people who live here take better care of things?" The inference of values based on the appearance of landscape can lead to other, problematic social judgments, especially when unanalyzed judgments cement unconscious biases based on class, race, age, and so on.

In response to the cultural, political, and urban upheavals that characterized the late 1960s and early 1970s, several seminal studies of landscape and values emerged in the late 1970s. Among these were *Changing Rural Landscapes* (1977), edited by interdisciplinary environmental researchers Ervin and Margaret Zube, which addressed recurring humanistic themes in the study of landscape change and focused on "analysis and interpretation of the landscape with particular attention to process, to human values, and to historical perspectives" (1977, xiii). At about the same time, cultural geographer Donald W. Meinig edited *The*

Interpretation of Ordinary Landscapes, a watershed in cultural landscape studies. His introductory essay, "The Beholding Eye: Ten Ways of Looking at the Landscape" (1979), foregrounds the importance of mental constructs and values in any viewer's landscape perceptions: "[E]ven though we gather together and look in the same direction at the same instant, we will not— we cannot—see the same landscape. We may certainly agree that we will see many of the same elements . . . but such facts take on meaning only through association; they must be fitted together according to some coherent body of ideas. Thus we confront the central problem: any landscape is composed not only of what lies before our eyes but what lies within our heads" (Meinig 1979, 33–34). Although Meinig does not use the term "values" here, they are surely subsumed in his reference to a "coherent body of ideas."

After 1980, the pace of production in cultural landscape studies accelerated. Much of the work in the 1980s and 1990s came from cultural geographers, although landscape architects, historians, architects, and ecologists contributed their shares. Reluctant to call it a "discipline" at first, some preferred to treat cultural landscape studies as a "rubric"—an unconsolidated project shared by a group of loosely allied writers from many fields. Today, however, the field of landscape studies has a distinct academic nomenclature and offers undergraduate majors in several colleges and universities.[2]

Nearly two decades after "The Beholding Eye," Paul Groth and Todd Bressi edited *Understanding Ordinary Landscapes* (1997). Building upon Meinig's edition, its purpose was to go further afield and survey "new directions in cultural landscape studies" (vii–viii).[3] Groth's introductory essay, "Frameworks for Cultural Landscape Study," along with a comprehensive bibliography, "Basic Works in Cultural Landscape Studies," present a still-valuable "capsule history" of the field. Groth writes, "The conviction among cultural landscape writers is that better knowledge of ordinary environments can foster deeper understanding of American people and American culture and can lessen the environmental dangers caused by people who cannot see and interpret their surroundings" (1997, 1–2). This conviction is just as valid today as it was in 1997, and motivates our work in this book.

LITERATURE FROM THE FIELD OF LANDSCAPE STUDIES

Thanks to J. B. Jackson (1909–1996), we know the etymology of the word "landscape" contains an astonishing range of values drawn both from language and practical life. In his magisterial essay "The Word Itself" (1984), Jackson writes, "We pull up the word by its Indo-European roots in an attempt to gain some insight into its basic meaning" (7). The word variously derives from: (1) *landschap* (Dutch origin)—referring to a site of productive work such as a farmstead or estate; (2) *landschaft* (German origin)—referring to a central corporate marker signaling a collective ideal of social identity, stewardship, and communal obligation; and (3) *landskip* (English origin)—referring to the visualization (painting or drawing) of territory. The tripartite heritage of "The Word Itself" helps explain the genesis of so many interrelated intellectual threads in cultural landscape studies.[4] From this, Jackson concludes: "*[L]andscape as a composition of man-made spaces on the land* is more significant than it first appears. . . . For it says that a landscape is not a natural feature of the environment but a *synthetic* space . . . functioning and evolving not according to natural laws but to serve a community. . . . A landscape is thus a space deliberately created to speed up or slow down the processes of nature" (original emphasis, Jackson 1984, 7–8).

While there is always a risk of oversimplifying the myriad interconnections of thought and purpose that inform the field of landscape studies, several broad categories of theory and method have each proven relevant for studying specific problems of landscape(s) and value(s). Below,

brief sketches of each problem area serve to introduce two or three classic works by seminal figures.[5] And if my selections seem idiosyncratic, then consider it a point of departure, like one of those quaint stacking signposts (the kind we can still see along backcountry roads in Maine or Ireland) for readers motivated to travel further along different paths.

Vernacular Landscape History

In addition to Watts, Meinig, and Groth, J. B. Jackson's influence on two generations of landscape scholars simply cannot be overestimated; indeed, most of the authors cited below acknowledge some intellectual debt to him. An essayist and thought leader, Jackson's role as editor of the journal *Landscape* was formative. Some of Jackson's late work was published as *A Sense of Place, a Sense of Time* (1994), and a retrospective collection of his essays, *Landscape in Sight: Looking at America* was published in 1997. One of Jackson's most influential and prolific students, John Stilgoe, has taught countless students and general readers alike. His *Common Landscape of America, 1580 to 1845* (1982) was followed by a succession of thematic studies on cultural landscapes. From *Metropolitan Corridor: Railroads and the American Scene* (1983) to *Train Time: Railroads and the Imminent Reshaping of the United States Landscape* (2007), Stilgoe has made broad and significant contributions to understanding vernacular and industrial typologies, as well as historical methods.[6]

Environmental Humanities

Humanistic disciplines such as environmental history, philosophy, and literature have examined questions of profound cultural values having to do with myth, imagination, and collective memory. Representatives of this genre include Clarence Glacken's *Traces on the Rhodian Shore: Nature and Culture in Western Thought from Ancient Times to the End of the Eighteenth Century* (1967), Robert Pogue Harrison's *Forests: The Shadow of Civilization* (1992), Lawrence Buell's *The Environmental Imagination* (1995), and Simon Schama's *Landscape and Memory* (1995). Challenging hybrids have also formed between the humanities and cultural geography, for instance Derek Gregory's *Geographical Imaginations* (1994). This is not even to mention many brilliant humanist nonfiction landscape writers like John McPhee and Annie Dillard.

Environmental/Ecological History

Beginning with Berkeley geographer Carl O. Sauer (1889–1975), environmental historians have understood that the shape and condition (morphology) of land and the human processes of making (and unmaking) it are thoroughly interconnected. Roderick Nash's *Wilderness and the American Mind* (1967) and *The Rights of Nature: A History of Environmental Ethics* (1989) are touchstones in this literature. Other notable works in this category include William Cronon's *Changes in the Land: Indians, Colonists, and the Ecology of New England* (1983) and his edited volume, *Uncommon Ground: Rethinking the Human Place in Nature* (1996). The landscape ecologist Joan Iverson Nassauer's edited volume *Placing Nature: Culture and Landscape Ecology* (1997) specifically addresses questions of landscape and values from an environmental perspective.[7]

Heritage and Historic Preservation

A scholarly field unto itself, the heritage literature is vast and complex; it has contributed many important concepts to the study of landscape and values, such as collective memory and intangible heritage. Heritage scholar David Lowenthal has powerfully influenced debates in landscape studies with *The Past Is a Foreign Country* (1985) and *The Heritage Crusade and the Spoils of History* (1998). Hybrid scholarship extends concepts from materialist heritage to the politics of landscape in *The Nature of Cultural Heritage and the Culture of Natural Heritage* (2006), coedited by David Lowenthal and the landscape geographer

Kenneth Olwig. Arnold Alanen and Robert Melnick's *Preserving Cultural Landscapes in America* (2000) applies the fundamentals of heritage planning to specific landscape problems.[8]

Cultural Materialism and Landscape

Geographers Kenneth Olwig and Don Mitchell stand as major representatives of this group. In *Landscape, Nature, and the Body Politic* (2002), Olwig traces the evolution of the landscape idea back to the seventeenth and eighteenth centuries when the emergence of the nation-state gave landscape much of its contemporary meaning. The following passage expresses his understanding of landscape as a condition resulting from the dynamics of cultural materialism: "Landscape is the expression of the practices of habitation through which the habitus of place is generated and laid down as custom and law upon the physical fabric of the land. . . . A landscape is thus a historical document containing evidence of a long process of interaction between society and its material environs" (Olwig 2002, 226). Olwig's focus on the communal and political dimensions of landscape expressed in aspects of social identity, cohesion, habitation, labor, justice, and institutions has inspired Don Mitchell and Tom Mels's essay in the present volume. Olwig and Mitchell have also edited a book entitled *Justice, Power and the Political Landscape* (2009) in which they seek to shift the understanding of landscape from a visual paradigm to one of "polity and place" (5).

Semiotics and Landscape Iconology

In the 1980s, poststructural critics and geographers began to combine aspects of visual and literary theory with cultural geography to reveal how values may be inscribed within built landscapes and public spaces. Seminal work in this group includes Denis Cosgrove's *Social Formation and Symbolic Landscape* (1984) and *The Iconography of Landscape,* edited by Cosgrove and Stephen Daniels (1988). W. J. T. Mitchell's edited work *Landscape and Power* (1994) firmly established the notion of landscape agency: "Landscape, we suggest, doesn't merely signify or symbolize power relations; it is an instrument of cultural power, perhaps even an agent of power that is (or frequently represents itself as) independent of human intentions" (1–2). Dimensions of agentic theory are painstakingly illustrated in *Landscapes of Privilege: The Politics of the Aesthetic in an American Suburb* (2004), a case study analyzed by James and Nancy Duncan over a twenty-year period to reveal and explain the nearly invisible infrastructure of visual and spatial codes that protect wealth and other exclusionary structures in the landscape.[9]

Phenomenology, Aesthetics, and Place Theory

Drawing from the mid-century philosophical traditions of Martin Heidegger and Gaston Bachelard, classic phenomenological theory easily cross-fertilized with landscape studies to explore place-based subjectivities. As distinct from "landscape," a geographical concept, the term "place" derives from philosophical or phenomenological concepts based on human responses to a range of variables acting within a defined space—memory, spatial enclosure or scale, personal association, sensory perception, and so on.[10] Conceptual relationships between values in and of *place* are explored in Edward Relph's classic *Place and Placelessness* (1976). Yi-Fu Tuan is another phenomenologist working in landscape geography, with his seminal work on place attachment, *Topophilia* (1974), and later *Passing Strange and Wonderful* (1993) on landscape aesthetics. Anne Spirn's *Language of Landscape* (1998) belongs to this group, although her work is so remarkably elastic it redraws new intellectual links between environmental history and humanities as well.[11]

Participatory and Place-based Design

Since the early 1990s, new patterns in the discourse on landscape and values have paralleled the rise of cultural pluralism. Landscape architects and community activists in particular have embraced this category of scholarship, none more comprehensively than Randolph Hester. Hester's concepts of community capacity and cohesion begin in a sustained relationship with a place. In his magnum opus, *Design for Ecological Democracy* (2006), Hester lays out fifteen principles for reforming landscape at a range of nested scales from the region to the household. "Enabling form" is didactic and supports the values of trust and deliberative democracy; "resilient form" is systematic and promotes self-adjusting natural and social systems; "impelling form" is inspiring and invites the values of hope and vision (8–9). Hester argues that only the deft integration of ecological sensitivity and democratic process will help us create resilient, enabling, and "joyful" communities—a true theory of values.[12]

In summary, and notwithstanding the fluid intellectual dynamics between the many scholarly camps in cultural landscape studies, each group has advanced specific problems and methods for interpreting, inhabiting, or constructing cultural landscapes. It remains, however, for design agents and practitioners, all those who construct and interpret landscapes, to put these theories to the test. The following section raises some general challenges that face all citizens of the landscape, while the tasks of setting professional priorities are discussed more fully in the concluding chapter of this volume.

Design: Values in Action

If values drive our landscape perceptions, then by extension they guide our design processes. If we accept the time-honored adage that *design* is a "course of action aimed at changing existing situations into preferred ones" (Steinitz 1995, 188), by definition, to design means *to add value*. To add value, in turn, requires a design agent (lawmakers, clients, landowners, designers, developers, managers, financiers, regulatory officials, and so forth) to apply a theory of value—that is, a theory of goodness. So it follows that all design operations require a working definition of values, whether implicit or explicit. Yet how many design agents can articulate, let alone defend, their values?

The word "values" is just as elastic as "landscape," encompassing quantitative, monetary, pragmatic, and moral benefit and/or worth, codes of behavior, and theories of goodness that have evolved through the centuries. Concepts of value are inherently philosophical, sandwiched between religion and morality (ideology) on the one hand and ethics (an applied operating system for making defensible choices) on the other. The meaning of the term "value" also may be redirected in various disciplinary contexts, including law, economics, philosophy, aesthetics, music, and mathematics. Quite beyond the associative and historical complexity of the meaning of "values," there are also contemporary relational, moral, and ethical issues to consider—all potential sources of landscape values and value judgments. So when the terms "landscape" and "values" are combined, the complexity of analysis multiplies exponentially.

As social constructions situated in specific contexts (class, education, gender, and so on), landscape values can never be naturalized or mistaken for self-evident or timeless truths. What is good for one person or species, at one scale, may be bad for another. In a retrospective study of how land use and landscape values have changed since the 1960s, Ervin Zube (1931–2001) outlined a framework to describe the mutually constitutive roles of landscape and designer/change agent. His model divided landscape actors into three groups: (1) humans as change agents generating (mostly negative) impacts on a passive landscape; (2) humans as passive receivers who regard the (scenic) landscape visually and emotionally; and (3) humans and landscape both actively engaged in interchange, feedback, and mutual transaction (1987, 38–39). The third "transactional" model describes a rich process of reciprocal (re)construction and "recognizes a full range of landscape experiences that can lead to the attribution of meaning and the valuing of specific landscapes" by designers and their clients (39). Valuing specific landscapes in turn shapes our identity, and thus our landscape ethics.

Because the recognition of landscape values and the responsibility to articulate, justify, and defend them are inherently political, such recognition touches upon "the often hidden agendas concerning polity and place we broach when we are ostensibly engaged in discourses concerning nature, landscape and the state of our environment" (Olwig 2002, xxxii). For example, let's imagine the construction of a new pipeline that conducts, say, Canadian shale gas to a port on the Gulf of Mexico. A variety of analysts have projected various results: jobs and personal wealth are (temporarily) created, energy supplies (temporarily) enhanced, habitat and groundwater systems (potentially, perhaps permanently) disturbed, landscape (physically and visually) marked and shaped at a continental scale, personal property compromised (or taken outright), and innovations for improving distributive infrastructure (possibly) discovered. In order to secure various forms and scales of "goodness," we might argue, there should be a system of values in place that orders our priorities and guides our decisions vis-à-vis shared landscape resources. But what shall we prioritize: diversity, historical significance, resilience, usefulness, profitability, social benefit, equity, beauty? How can we ever hope to identify a contemporary calculus providing "the greatest good for the greatest number," as Jeremy Bentham (1748–1832) had once attempted?[13]

To have any meaning at all, any theory of goodness that guides the development of ethical systems (by virtue of which we are able to make these kinds of decisions at all) must be part of a web-work of socially constructed and plural value systems that have been politically contested and resolved. Some good examples exist in environmental legislation, for instance, design standards supporting many forms of public accessibility, as well as new-again design concepts such as aging-in-place and multifunctional streets. However, many environmental design agents with enormous impact on built landscapes rarely have the opportunity or the inclination to examine or debate such theories. Worse, many citizens are either so detached from landscape processes or so ignorant of the spectrum of *possible* alternative values, they are incapable of exercising any influence over the forms of environmental products typically on offer (housing, streets and parking patterns, public water supplies, landfills, energy supplies, agriculture, and so on).

Nevertheless—for better or worse—values can and do change, and the inevitable conflicts that result also find their way into built landscapes. Values explain the past and presage the world yet to come. Environmental design requires sensitive mediation between value sets operating at different scales of time and impact, and it benefits from participatory engagement, imaginative projections, and/ or pre-emptive responses to unanticipated impacts.

Contrasting the idea of anticipatory design with performance measurement, landscape planner and theorist Simon Swaffield shows they are just two sides of the coin we call "values": "As De Bono (2000, p. 218) puts it, 'Design thinking is very different from traditional judgment thinking. For judgment thinking, the desired output is truth or apparent truth. For design thinking, the output is value. . . .' Indeed it can be argued that the values being sought or created through [design] are not variables within either design or science, but express the cultural norms and social institutions within which design and science are undertaken, and the values recognised in particular places and landscapes by their communities. Values thus frame the processes of design, management and science as well as the choices they involve, and are embodied in design outcomes" (Swaffield 2013, 2). Swaffield's point, that even the most rigorous "judgment thinking" is subject to the hidden influence of values, is absolutely essential. One need only consider the historical role of values in making "objective" performance calculations, or the selection of case studies that represent any canon (whether in law, medicine, religion, or design). Solving problems is admirable, but to be solvable, every problem must first be framed by an intellectual question—again, begging an articulate design theory of values.

How Landscape Values Shape Culture

Ever since the Vitruvian triad (firmness, commodity, delight), design theory has comprised a theory of values. Every theory of form and composition is a theory of values—for example, the anthropomorphism of neoclassicism; the streamlined industrial shapes of mid-century modernism. Every theory of performance or function is a theory of values—for example, emancipation from social myths revealed in semiotics; the inadequacy of single disciplines implied in landscape urbanism. Whether concerned with an idealized Renaissance symmetry thought to reflect the mind of God, or the resilience, distribution, and abundance of pollinator species, the intention of design theory is to argue and secure various aspects of a greater good.

Taking this further, it is possible to "read" values expressed in landscape design—for

Fig. 1.2. Chicago skyline seen beyond the Lurie Garden in Millennium Park. Lurie Garden designed by Gustafson, Guthrie, and Nichols, with Piet Oudolf, 2005. Photograph courtesy of Mary Pat Mattson, 2014.

example, the 2.5-acre Lurie Garden, a part of Chicago's Millennium Park (fig. 1.2). One of the world's largest roof gardens, Millennium Park was built using air rights over the old Illinois Central railyards (now with vast underground parking lots serving visitors to downtown). The garden's materials (plants, water, innovative surfacing) have been selected to recall ancient prairie ecosystems (99.99 percent destroyed) and to present a virtuous spectacle of Chicago's spectacular

waterfront real estate juxtaposed against a hybrid "Nature." As an innovative public-private development project, the value(s) of the Lurie Garden and Millennium Park speak to several theories of "goodness": replacing degraded industrial property with functional public space, supporting cultural institutions with publically funded infrastructure (Art Institute of Chicago, Chicago Symphony Orchestra, others), and providing public space for private benefit (luxury high-rise towers). Meanwhile, heavy social and arts programming in the park has returned clear profits to Chicago in terms of increasing real estate values, tourism and shopping dollars, and a symbolic front door that has renovated the image of the city.

Not all expressions of value are quite so innovative. For example, consider how many competing values are invoked by the perfectly ordinary decision to plant street trees in raised beds as opposed to grates: healthy root zone for trees versus convenient movement space for people; higher construction costs versus higher carbon footprint; greater tripping hazards and maintenance costs versus higher expected tree mortality; more carbon sequestration and surface cooling versus lower visibility for street facades and store windows, and so on. Trade-offs involved in such alternatives offer different benefits to different constituencies, yet in effecting many ordinary design decisions, the code of values is not typically debated—not by clients, designers, users, or public administrators. True: given the complexity of hundreds of design decisions required by the average environmental design project, to debate every point seems practically absurd. On the other hand, if developers and designers argued for *values first*, letting program, form, and material follow, perhaps they could better explain and defend their choices.

The best explanation for how landscape values shape culture comes from the theory of *cultural materialism,* developed in the early 1970s by British cultural historian and literary critic Raymond Williams (1921–1988). By extending principles of cultural materialism to landscapes, we can see how closely landscape practices and values are linked to broader mechanisms that drive cultural change. The term "dominant culture" generally refers to "norms" (including values) that develop in societies over time. Ubiquitous landscape patterns and institutions (for example, factory farms, gated communities, shopping malls) tell of *dominant cultural values*—the value of time, space, land, labor, people, things, and the relationships between all of these.[14] According to Williams, pressures exerted by dominant culture (for example, American-style capitalist democracy) are inescapable. Yet dominant culture is by no means static or monolithic; it ensures its own vitality and relevance by engaging competing subcultures in a process called "incorporation"; in so doing, both subordinate and dominant cultural values are gradually transformed. Dominant culture is thus Janus-like, looking in two directions—backward toward *residual values and practices* and forward to *emergent values and practices.*

Residual values and practices are remnants of an older society, perhaps receding to the sidelines of history or serving still as enduring symbols of origin myths. We can read residual values in the patterns of living historical landscapes or vernacular traditional communities—for instance, Amish farm communities or the central common or green of a New England village. Even when dominant culture moves away from residual values, it cannot resist a backward glance for orientation. Residual values are thus typically tolerated and maintained in a condition of stasis, neither flourishing nor waning away completely, as they remain symbolically powerful markers of authenticity and legitimacy.

Emergent values and practices belong to anticipatory technical, formal, and social innovation (for example, the cultural or scientific avant-garde). Because emergent values identify and pursue new responses to changing needs and capacities, contemporary culture is typically eager to incorporate emergent cultural values. Examples of incorporated values that have influenced the shape of the landscape include sustainability, multiculturalism, fuel efficiency, personal mobile communication, and so on. According to Williams, dominant culture needs to incorporate newness "if it is still to be felt as in real ways central in all our many activities and interests" (1973, 14). Emergent values leave their mark on the landscape in a variety of forms, for instance in newly recentralized settlement patterns (consider the gentrification of Brooklyn) and interactive spatial programming (such as Theater Square in Rotterdam, designed by West 8, or networked car-sharing strategies such as ZipCar). Readers can undoubtedly identify dozens of other examples in their local landscape.

Williams identifies two other cultural processes that (for the most part) exist off the trajectory of dominant culture: *alternative values and practices*, and *oppositional values and practices*. Alternative values and practices exist parallel to, but are generally aloof from, dominant culture. Because alternative values tend to be highly specialized, idiosyncratic, or autonomous, they can seem somewhat less attractive for mass commercial or institutional incorporation. On the other hand, oppositional values are clearly defined and oriented by a codependent, if antagonistic, relationship with dominant culture—both in their memes and their methods. After starting out as marginal, opposition may be gradually incorporated and embraced by the general public (consumers, voters, pop culture), and/or legitimized by the institutional superstructure (art museums, universities). Over time, oppositional themes may be officially assimilated through constructive legal compromise (civil rights, environmental regulation) and/or modification of the original aims (hip hop), sometimes to the

point where opposition is muted or becomes an alternative lifestyle choice (punk culture, queer culture). Interestingly, Williams points out that both alternative and oppositional values may continue to adapt *within* emergent or residual culture.[15]

In the four decades since 1973, the dynamic elasticity of Williams's cultural worldview seems to have accommodated the long sequence of cultural transformations that adult generations worldwide have witnessed. Given a clearer understanding of this patient view of cultural values and processes, landscape architects may be able to seize opportunities that are simultaneously moral, intellectual, and practical. In particular, because so many emergent values are eventually incorporated by dominant culture, alternative-emergent and oppositional-emergent practices and values represent the thin edge of the wedge of the agency of artists, designers, teachers, and other creative activists. And as we have seen in Millennium Park (fig. 1.3), intensifying the

Fig. 1.3. Anish Kapoor, *Cloud Gate* (2004), signature public sculpture in Millennium Park, Chicago. Photograph courtesy of Tanvi Shah, 2014.

development and incorporation of emergent values in landscape practices is a potentially powerful strategy to effect broader societal and environmental change.

The Essays

The various studies and positions presented in this book represent a complex survey of values in action, operating at different scales of time, place, and impact. Indeed, these essays do two things: each describes a new problem of value(s) that has emerged along with new technology, capital, or theory, and each investigates value(s) in new ways, at new scales, with new methods. Further, these studies illuminate the evolutionary processes of cultural materialism, and animate the roles of landscape values and practices in reshaping places as well as institutions.

AESTHETICS, PRAGMATICS, DESIGN

It is no accident that this volume begins with two strong philosophical arguments. In "Beyond Sustaining Beauty: Musings on a Manifesto," Elizabeth Meyer presents a sweeping meditation upon her provocative earlier work "Sustaining Beauty: The Performance of Appearance" (2008). Having argued earlier for the reintroduction of aesthetic awareness into the so-called "Three

Es" of sustainability (environment, economy, and equity), she touched a chord in the value system of landscape architecture that resonated far beyond her initial expectations. In this reasoned and intellectually courageous response to subsequent positive and negative feedback, Meyer traces the new turns that her thinking *and her values* have taken. By reconsidering the role of beauty in the environment, and in our experience of life and place, her hope is that humanity's "egocentric" worldview may shift to "a more bio-centric perspective." Rather than conventional aesthetic codes of sweetness and perfection, however, Meyer envisions a new conception of sustainable aesthetics that may, exactly like good art, "challenge through difference and dissonance" (32).

In "The Value of Values," Kathryn Moore defines aesthetics not as "what something looks like" or a form of "cultural service" but as nothing short of the entire social, physical, and cultural context of our lives (54). From Moore's opening salvo—"A radical redefinition of the relationship between the senses and intelligence is way overdue"—she turns received wisdom on its head, constructing a new philosophical argument to systematically challenge the notion of a sensory interface or sensory mode of thinking. The radical premise of Moore's work is that rather than argue "we should recognize the intelligence of perception, it is

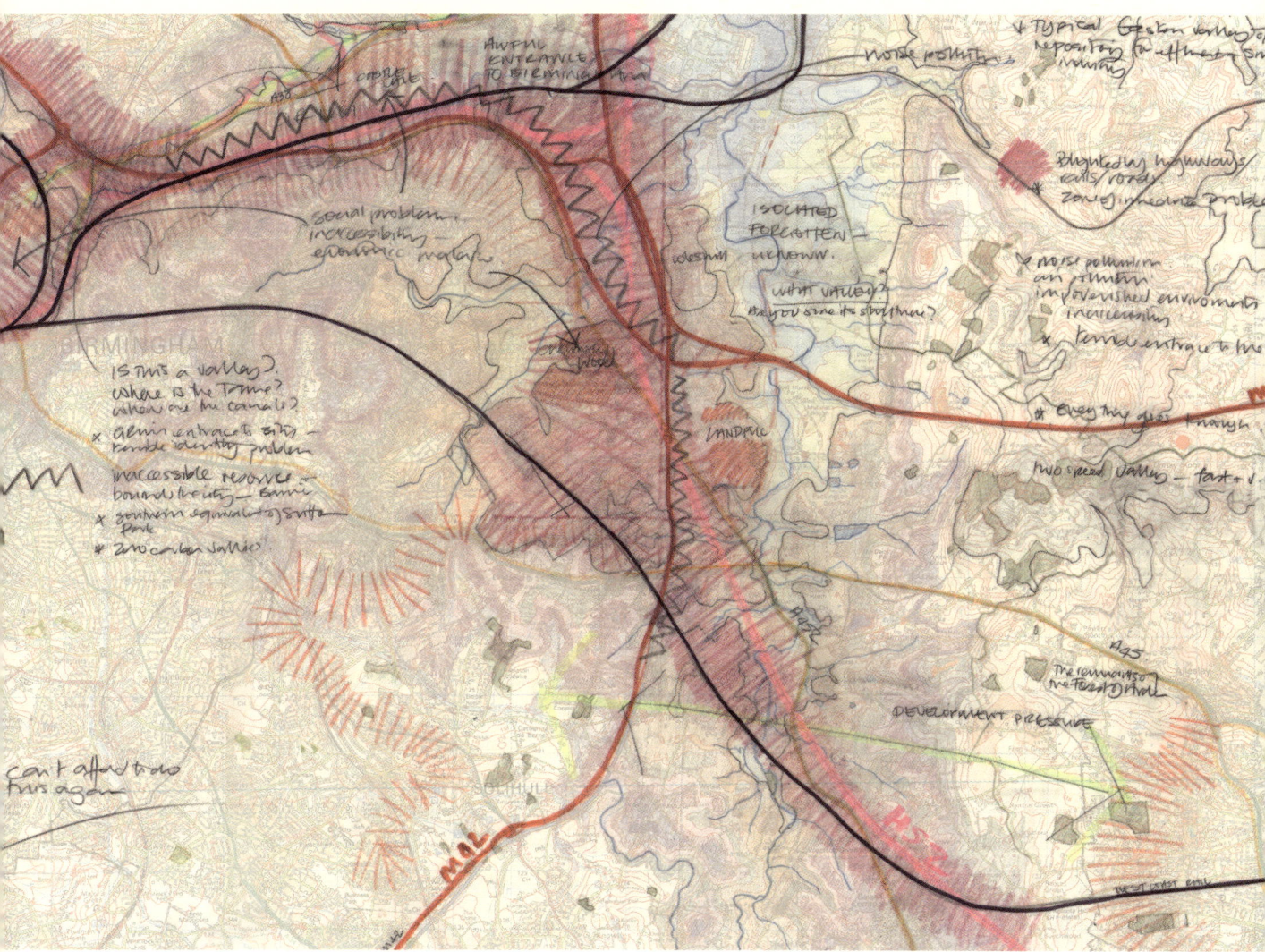

better to argue that perception *is* intelligence" (62). Moore contests the conventional bias towards objective or "scientific" modes of thought, institutionalized since the Enlightenment. With a slight shift in perspective, she argues, conventional divides (facts and values, nature and culture, senses and intelligence, language and emotion) can be reassembled to fundamentally redefine the nature of design expertise. Based on a pragmatist philosophy that offers a middle way through such impossible oppositions, Moore demonstrates through her own teaching and design practices (fig. 1.4)

Fig. 1.4. Detail of working composite drawings for the HS2LV master plan with value-laden annotations, 2012. Designed and drawn by Kathryn Moore. Used by permission.

how a new approach with "wider social and political implications" may be applied (67). The ability to articulate why things look like they do, or understand why landscapes are considered beautiful and meaningful, she argues, is a vital aspect of design expertise that can be taught and learned over time.

THE ETHICS OF INTERVENTION

The next two essays explore the social and environmental ethical dilemmas set in motion by unintended consequences of unexamined hubris, especially if coupled with well-intentioned zeal. "De-domestication and the Wild," by Catherine Seavitt Nordenson, is a fascinating story of genetic manipulation and landscape management in Europe at a scale and time difficult to comprehend. Sketching the history of Nazi ambitions for back-breeding large prehistoric European herbivores for the purpose of landscape management, Seavitt Nordenson raises uncomfortable questions about current and proposed landscape management programs. What is the fate of these genetically modified but essentially still-domestic animals left to fend for themselves in an open-ended ecosystem? Anticipating the inevitable dynamics of the predator-prey food chain does not require a dramatic leap of imagination. In such cases, what is the moral or "humane" responsibility of the "expert" in engineering alternative and unpredictable outcomes? In the hierarchy of values expressed in these Dutch landscape management regimes, which species should have supremacy over the others?

Kyle Brown shares a seasoned perspective on the roles and responsibilities of those in higher education who may seek to assist marginalized communities. Director of the Lyle Center for Regenerative Studies (a sixteen-acre laboratory for teaching and research on sustainability at the California State Polytechnic University, Pomona), Brown urges educators and other community-minded activists to cultivate a critical social awareness of the unintended consequences sometimes set in motion by service-learning projects and similar studies. "Toward Ecological Sovereignty: The Regenerative Communities Initiative" explores the pitfalls of academic expertise and its reciprocal dependency, or "clienthood," which often mirrors the same asymmetrical power relations that exist between elite professionals and marginalized communities. In outlining better pathways to ecological sovereignty, Brown points out, a community may be empowered in other ways—socially, politically, and intellectually. And by awakening the community's awareness of their own place-based values and letting them take charge, he argues, both environmental perception and political self-determination stand a better chance of taking root.

COMMEMORATION, COMPROMISE, AND CRITIQUE

What value(s) do we place on public memory? The next two essays offer insightful readings of the way America remembers its dead and its survivors, and show us how private emotions may become a proxy in negotiating larger social, political, and national agendas. In "Memory Work: The Submissions to the Oklahoma City Memorial Competition," Martin Holland presents a close reading of the design competition entries to commemorate the April 1995 attack on the Murrah Federal Building in Oklahoma City. The dust had barely settled when programming for a memorial began, a defiant gesture in the face of overwhelming loss that served citizens as a kind of collective public therapy. By grouping, interpreting, and naming dominant themes in many non-premiated submissions, Holland profiles the mood of the nation. However, through the "tyranny of the therapeutic" (111), bringing real and latent chaos under spatial and symbolic control and avoiding images of risk, confrontation, or trauma, Holland argues

that the "forgotten" design submissions occlude the essential political lessons of the Oklahoma City bombing.

A nearly simultaneous event, the 1995 dedication of the Korean War Veterans Memorial on the National Mall in Washington, D.C., expresses a different kind of agenda; rather than grief masked by tranquility, the memorial cloaks perpetual geopolitical struggle in the glorification of military service. Alan London presents a meticulously investigated, blow-by-blow account of how design decisions were compromised by a larger, nationalist agenda. In "Honoring Korean War Veterans: Conflicting Values of Commemoration," London sets out a concise history of the changing national attitude towards memorials and monuments as a timeline of political values and their reception. In part, he argues, the figural literalism of the Korean War Veterans Memorial is a reaction by a generation of vets to the austere minimalism of the Vietnam Veterans Memorial—called by its critics the "black gash of shame." Were we simply seeing the slow pendulum of commemoration styles swing from abstraction back to narrative? If the value of an explicit curatorial narrative in "memory work" is far from clear, in both Holland's and London's essays the agency of the designer in relation to "multiple publics" is even less so.

THE VALUE OF WORK IN CULTURAL LANDSCAPE

Two essays present case studies in landscape geography that fit a more classic mold. In his insightful history "Forging Commonplace: The Old Northwest Territory's Emblematic Transition from Wilderness to Landscape," Stephen Sears argues that the determined labor of generations gradually transformed the American Midwest. From the myth of an American Eden to the productive landscape of the promised land, the ethics of work shaped (and still define) the land and its culture. Sears details how this conceptual shift helped construct early American values in a powerful and transformative way. The reduction and consolidation of wet and dry prairies, forests and mountains, rivers and bluffs into "landscape," Sears explains, was one of the most important measures— literally and metaphorically—of the young country's progress, as well as the crucible for its emerging value system. During the mid-twentieth century, when the tangible legacy of the American Eden was all but lost, the myth lived on even more powerfully in fiction and images that continue to shape American narratives to this day.

Themes of landscape myth and iconology form an obvious nexus between Sears's "Commonplace" and Jennifer Britton's "Imbibing *Terroir*: Values in Napa Valley's Cultural Landscape of Wine." Using similar methods of image analysis in a different geographic context (Midwest vs. northern California), Britton tells a story of myth-making and internal contradictions aimed to "transcend reality in pursuit of an ideal place" (157). In both Britton's and Sears's essays, labor's presence, as well as its absence, is key to transforming both land and image. Just as winemaking renders mineral soil and atmospheric conditions into the subtle notes of *terroir,* we may parse images into particular nuances of meaning. Britton uses semiotic analysis to show how the politics and patterns of place—such as invisible labor, chemical enhancements, social displacement, and global capital—may undermine the "romantic image" of California's wine-growing regions and expose the mythmaking machinery of the wine industry.

POWER RELATIONS IN GLOBAL URBAN LANDSCAPE

Flowing methodologically from the preceding essay, work by Amita Sinha and Rajat Kant explores the uses and abuses of public mythmaking in support of powerful political agendas. "City of Nawabs to City of Elephants" describes a remarkably ambitious landscape construction program undertaken in the city of Lucknow, India, by the populist Mayawati, four-term chief minister of the state of Uttar Pradesh. Constructing a series of monumental public spaces served her rule as a means of securing public admiration, empowering her own ruling political party, and consolidating her main base of political support. Although the authors are critical of the ways the spaces function on the neighborhood level, Mayawati's public landscapes do serve as major tourist attractions—as *spectacle*—precisely the way the term is defined by Guy Debord (1995). If we understand spectacle to mean capital accumulated to the point where it mediates and replaces authentic social interactions with well-choreographed distractions and awe, then one point is evident: agendas of both the left and the right may be served equally by the keen deployment of spectacular iconography.

Don Mitchell and Tom Mels also study the impact of concentrated political and economic capital on the disposition of landscape resources. In Gotland, Sweden, and Youngstown, Ohio, industrial landscapes are "the hard surfaces of life," a lived index of the impacts of finance capital (209). From Scandinavia to the American Rust Belt, Mitchell and Mels span the globe in a powerfully argued reflection on capital, landscape resources, and justice. We learn that both investment and disinvestment of industrial capital cause equally traumatic changes on the landscape—changes to its visual character, its structure, its ecology, its

sense of community, and integrated power relationships. Mitchell and Mels thus present a larger, more fundamental consideration of social landscape ecology and the dynamic cycles of *value* over time. Whether industrial farming or manufacturing, whether corporate or locally owned, landscape today can hardly avoid the ineluctable logic of global urban capital systems. Within this system, if justice is not maintained as a strong priority, it becomes increasingly difficult for landscape architects to find any center of balance.

Conclusion

Some of the essays collected here present case studies on landscape values in order to test and challenge accepted theories of goodness (Seavitt Nordenson, Holland, London, Sears, Sinha and Kant). Others propose a specific theory of values illustrated by exemplary cases (Meyer, Moore, Brown, Britton, Mitchell and Mels). But all the contributors lead the reader on a journey of discovery to witness how different belief systems and approaches to practice not only shape values into landscapes but also influence community hopes, fears, aspirations, and capacity for growth.

These essays make readers aware that both innovations and transgressions of landscape values (alternative/emergent or oppositional/emergent practices) can engender the creation of a vivid and unmistakable sense of place, and that new landscape practices may challenge and force our reexamination of dearly held values. Whether we choose to uphold or resist established values or create entirely new kinds of places for alternative values to emerge, all of us are agents and shapers of landscape(s). And as critical participants in all kinds of value(s)-based practices, we are therefore accountable for how our values materialize—through our policies, our designs, our choices, and especially our patterns of expectation.

NOTES

1. An abbreviated early version of this essay (Deming 2013) was delivered at a symposium at Tsinghua University, Beijing. The author is grateful for constructive feedback from participants, especially Anne Spirn and Richard Weller.

2. Examples are numerous: Harvard's Research Library at Dumbarton Oaks, Washington, DC, specializes in garden and landscape studies; Smith College was the first to offer an undergraduate degree program in landscape studies; students at Cornell University and the University of Illinois, Urbana-Champaign, may choose undergraduate minors in landscape studies; and both the University of Illinois and Indiana University offer a doctoral minor called landscape studies. The Society for Landscape Studies (UK), founded in 1979, publishes the journal *Landscape History*; the Foundation for Landscape Studies (New York) is a public not-for-profit charity directed by Elizabeth Barlow Rogers, and so on.

3. Groth and Bressi's long list of contributors overlaps only slightly with Meinig's (J. B. Jackson and David Lowenthal), and includes broader participation from landscape architects, architects, landscape and social historians, and visual studies scholars.

4. None of these threads or their textual communities are particularly antagonistic; rather, they represent methodological alternatives. Authors in various camps acknowledge, parallel, and cross-reference each other liberally; for instance, Yi-Fu Tuan writes the foreword for Olwig's *Body Politic* (2002), James Corner critiques and advances Cosgrove's core theses, and so on.

5. Another editor generating a list serving the same purpose would likely feature many of the same authors and titles, along with unique selections of their own. A citation search for any of the authors mentioned will yield a comprehensive list of current books and articles by others.

6. Stephen Sears's essay "Forging Commonplace" (this volume) connects vernacular landscape interpretation with the next category—environmental humanities.

7. Catherine Seavitt Nordenson's essay "De-domestication and the Wild" (this volume) links the category of ecological history with human history.

8. "City of Nawabs to City of Elephants," contributed by Amita Sinha and Rajat Kant (this volume) belongs to the heritage category with links to semiotics and participatory planning.

9. Martin Holland, Alan London, and Jennifer Britton each use semiotic methods in their respective studies (this volume); Britton also touches on cultural materialism, while London and Holland link to heritage studies.

10. Landscape and place are intellectually distinct ideas: a specific landscape certainly can be a place, but then so can a soup kitchen or the foyer of the Paris Opera. Unless it is meant metaphorically, however, it

cannot as easily be said that the soup kitchen or the foyer of the Paris Opera is a landscape.

11. In their respective essays "Beyond Sustaining Beauty" and "The Value of Values" (this volume), both Elizabeth Meyer and Kathryn Moore challenge traditional phenomenology and aesthetic theory.

12. Kyle Brown's essay on "Ecological Sovereignty" (this volume) pursues this line of cultural landscape value and process.

13. Jeremy Bentham, inventor of the Panopticon, also founded utilitarianism, a liberal utopian philosophy important to modern legal philosophy.

14. Contemporary doctrines of multiculturalism reject the idea of uncontested dominance of one social group over another and argue in favor of greater cultural equity. Nevertheless, cultural identity and values expressed by individual members of diverse social groups usually remain fluid and situational and are typically inflected by dominant culture at some level.

15. For instance, *alternative-residual* cultures might include fundamentalist utopian or religious communities; *alternative-emergent* cultures might look like collective eco-housing societies living "off-grid" and developing their own high-tech systems for food and energy production, do-it-yourself culture, or new Internet-based "sharing economies" and bit-coin currency. One may see *oppositional-residual* landscape cultures expressed in radical militia compounds or aggressive sectarian societies. One may also see *oppositional-emergent* values in new forms of Internet resistance such as hacking, critical social networks or blogs, slow food, in cultures of creative dissent such as skateboarders or body art, and so on.

REFERENCES

Alanen, Arnold, and Robert Melnick. 2000. *Preserving Cultural Landscapes in America.* Baltimore: Johns Hopkins University Press.

Buell, Lawrence. 1995. *The Environmental Imagination: Thoreau, Nature Writing, and the Formation of American Culture.* Cambridge, MA: Harvard University Press.

Cosgrove, Denis. 1984. *Social Formation and Symbolic Landscape.* Madison: University of Wisconsin Press.

Cosgrove, Denis, and Stephen Daniels, eds. 1989. *The Iconography of Landscape.* Cambridge, UK: Cambridge University Press.

Cronon, William. 1983. *Changes in the Land: Indians, Colonists, and the Ecology of New England.* New York: Hill & Wang.

———, ed. 1996. *Uncommon Ground: Rethinking the Human Place in Nature.* New York: W. W. Norton & Co.

de Bono, Edward. 2000. *New Thinking for the New Millennium.* New York: Penguin Books.

Debord, Guy. 1995. *The Society of the Spectacle.* Trans. Donald Nicholson-Smith. New York: Zone Books.

Deming, M. Elen. 2013. "Tomorrow's Values in Landscape Architecture Today." In *Landscape Architecture of Tomorrow: International Forum Proceedings,* 3–11. Beijing: Tsinghua University School of Architecture and the Chinese Society of Landscape Architecture.

Duncan, James S., and Nancy G. Duncan. 2004. *Landscapes of Privilege: The Politics of the Aesthetic in an American Suburb.* New York: Routledge.

Frost, Robert. 1964. "Mending Wall," *Complete Poems of Robert Frost,* 47–48. New York: Holt Rinehart and Winston, Inc.

Glacken, Clarence. 1967. *Traces on the Rhodian Shore: Nature and Culture in Western Thought from Ancient Times to the End of the Eighteenth Century.* Berkeley: University of California Press (1976 reprint).

Gregory, Derek. 1994. *Geographical Imaginations.* Cambridge, MA: Blackwell.

Groth, Paul, and Chris Wilson, eds. 2003. *Everyday America: Cultural Landscape Studies after JB. Jackson.* Berkeley: University of California Press.

Groth, Paul, and Todd W. Bressi, eds. 1997. *Understanding Ordinary Landscapes.* New Haven, CT: Yale University Press.

Harrison, Robert Pogue. 1992. *Forests: The Shadow of Civilization.* Chicago: University of Chicago Press.

Hester, Randolph T. 2006. *Design for Ecological Democracy.* Cambridge, MA: MIT Press.

Jackson, John Brinkerhoff. 1984. "The Word Itself." In *Discovering the Vernacular Landscape.* New Haven, CT: Yale University Press, 2–8.

———. 1994. *A Sense of Place, A Sense of Time.* New Haven, CT: Yale University Press.

———. 1997/2000. *Landscape in Sight: Looking at America.* Ed. Helen L. Horowitz. New Haven, CT: Yale University Press.

Lewis, Peirce. 1979. "Axioms for Reading the Landscape." In *The Interpretation of Ordinary Landscapes: Geographical Essays,* ed. D. W. Meinig, 11–32. New York: Oxford University Press.

Lowenthal, David. 1985. *The Past Is a Foreign Country.* Cambridge, UK: Cambridge University Press.

———. 1998. *The Heritage Crusade and the Spoils of History.* Cambridge, UK: Cambridge University Press.

Meinig, Donald W. 1979. "The Beholding Eye: Ten Versions of the Same Scene." In *The Interpretation of Ordinary Landscapes: Geographical Essays,* ed. D. W. Meinig, 33–48. New York: Oxford University Press.

Meyer, Elizabeth K. 2008. "Sustaining Beauty: The Performance of Appearance—A Manifesto in Three Parts." *Journal of Landscape Architecture* 3 (Spring): 6–23.

Mitchell, W. J. T., ed. 1994. *Landscape and Power.* Chicago: University of Chicago Press.

Nash, Roderick Frazier. 1967/2001. *Wilderness and the American Mind.* 4th ed. New Haven, CT: Yale University Press.

———. 1989. *The Rights of Nature: A History of Environmental Ethics.* Madison: University of Wisconsin Press.

Nassauer, Joan Iverson, ed. 1997. *Placing Nature: Culture and Landscape Ecology.* Washington, DC: Island Press.

Olwig, Kenneth Robert. 2002. *Landscape, Nature and the Body Politic: From Britain's Renaissance to America's New World.* Madison: University of Wisconsin Press.

Olwig, Kenneth Robert, and David Lowenthal, eds. 2006. *The Nature of Cultural Heritage and the Culture of Natural Heritage.* New York: Routledge.

Olwig, Kenneth Robert, and Don Mitchell, eds. 2009. *Justice, Power and the Political Landscape.* New York: Routledge.

Relph, Edward. 1976/1983. *Place and Placelessness.* London: Pion Ltd.

Rosenberg, Elissa, ed. 1993. *Design & Values: CELA Conference Proceedings,* vol. 4. Charlottesville: University of Virginia.

Schama, Simon. 1995. *Landscape and Memory.* New York: Alfred A. Knopf.

Spirn, Anne Whiston. 1998. *The Language of Landscape.* New Haven, CT: Yale University Press.

Steinitz, Carl. 1995. "Design Is a Verb; Design Is a Noun." *Landscape Journal* 14 (Fall): 188–200.

Stilgoe, John R. 1982. *Common Landscape of America, 1580 to 1845.* New Haven, CT: Yale University Press.

———. 1983. *Metropolitan Corridor: Railroads and the American Scene.* New Haven, CT: Yale University Press.

———. 2007. *Train Time: Railroads and the Imminent Reshaping of the United States Landscape.* Charlottesville: University of Virginia Press.

Swaffield, Simon. 2013. "Empowering Landscape Ecology—Connecting Science to Governance through Design Values." *Landscape Ecology* 28(6): 1193–1201, link.springer.com/article/10.1007/s10980-012-9765-9#page-1 (accessed May 2013).

Tuan, Yi-Fu. 1974. *Topophilia: A Study of Environmental Perception, Attitudes and Values.* Englewood Cliffs, NJ: Prentice-Hall.

———. 1993. *Passing Strange and Wonderful: Aesthetics, Nature and Culture.* Washington DC: Island Press.

Watts, May Theilgaard. 1975. *Reading the Landscape of America,* revised ed. New York: Collier Books.

Williams, Raymond. 1973. "Base and Superstructure in Marxist Cultural Theory." *New Left Review* I/82 (November/December): 3–16. Reprinted in *Culture and Materialism: Selected Essays,* 31–49. London: Verso, 1980.

Zube, Ervin H. 1987. "Perceived Land Use Patterns and Landscape Values." *Landscape Ecology* 1(1): 37–45.

Zube, Ervin H., and Margaret J. Zube, eds. 1977. *Changing Rural Landscapes.* Amherst: University of Massachusetts Press.

ELIZABETH K. MEYER

Beyond "Sustaining Beauty"

MUSINGS ON A MANIFESTO

In 2007, I had grown dissatisfied with the status of sustainable design discourse in landscape architecture. By adding an additional dimension to the "three Es" of sustainability—environment, economy, and equity—I asserted that sustainable design needed to be more than a technical response to improving ecosystem function or increasing access to public space. From Vitruvius to modern landscape architects Charles Eliot and Jens Jensen, designers have written of the inextricability of aesthetics, function, and structure. Motivated by that knowledge, I called for (re)inserting aesthetics into the sustainability triad (see list below).[1]

After writing this declaration of my values and beliefs in manifesto form, I continued to expand my knowledge of sustainability discourse in readings and discussions with my students, as well as through public lectures.[2] Two journals, one academic and one professional, published the manifesto, thus increasing the range and type of reader (Meyer 2008a and 2008b).[3] Their collective responses were startling. In the twenty-five years I have been lecturing and writing about contemporary landscape design topics, no other publication has touched such a strong chord, elicited so many varied responses, or reached such a broad audience.

Therefore I want to share my musings about the values, often implicit, behind the reception of "Sustaining Beauty," as well as my own unexamined assumptions. I want to renew my investment in the entanglement of social aesthetics with experiments in making and living. Finally, I also want to lay bare my process of reflection, critique, and extension —the invisible work between publications that occurs in conversations and debates with students, colleagues, and strangers.[4]

BASIC TENETS FROM THE MANIFESTO "SUSTAINING BEAUTY"

1. Sustaining culture through landscape (Landscape architecture is a cultural practice, not just a professional practice)
2. Cultivating hybrids—the language of landscape[5]
3. Beyond ecological performance
4. Natural process over natural form
5. Hypernature—the recognition of art[6]
6. The performance of beauty[7]
7. Sustainable design = constructing experiences
8. Sustainable beauty is particular, not generic[8]
9. Sustainable beauty is dynamic, not static
10. Enduring beauty is resilient and regenerative
11. Landscape agency—from experience to sustainable praxis.[9] (Meyer 2008a and 2008b)

Background: Reception of "Sustaining Beauty"

"Sustaining Beauty" was intended to provoke and challenge the status quo as much as to persuade. A design manifesto requires the declaration of beliefs, the taking of a stand. Disgruntled as I was with what landscape architects were saying, or not saying, about sustainability, the manifesto seemed to be the most optimistic and efficacious mode of redirecting the discourse.

Sustainable landscape design is generally understood in relation to three principles—ecological health, social justice, and economic prosperity. Until recently, aesthetic values had not factored into the discourse of sustainability, except in negative asides conflating vision, visuality, and formalism with the aesthetic, and rendering all of that secondary to ecological function, performativity, quantitative metrics, or ecosystem services. To the contrary, however, I believe that, as a body of knowledge and a way of experiencing the world, social aesthetics can play a critical role in a sustainability agenda. It will take more than ecologically regenerative designs for our early twenty-first-century neoliberal consumer culture to be sustainable.

What is needed are designed landscapes that provoke those who experience them to be more aware of how their actions affect the environment, and to care enough to make changes in their actions. This involves recognition of the role of aesthetic environmental experiences, such as beauty, wonder, awe, ugliness, and repulsion, in re-centering human consciousness from an egocentric to a more biocentric perspective. Such recognition is dependent on new conceptions of human and nonhuman entanglements.

Many who heard or read my manifesto agreed that landscape architects were too narrowly construing sustainability as a new form of extreme functionalism, realized through green infrastructure or eco-technologies. But they did not share my understanding of aesthetics. Some thought of aesthetics as a retrograde concern, or a call for universal types, or a synonym for natural beauty.[10] Post-publication, therefore, I started revising the manifesto, finding additional sources for my assertions, combining a couple of the tenets, and adding a new one called "Sustainable aesthetics challenge through difference and dissonance." This tenet was necessary to shake up those who could not get beyond a limited sense of beauty, so they might appreciate a broader argument about the role of new, alternative forms of aesthetic experience in a sustainable design agenda.

The questions about these issues compelled me to read philosophers such as Arthur Danto (1999), Alexander Nehamas (2007), and Elaine Scarry (1999) more closely. More significantly, other theoretical arenas opened up, such as the theories of affects.[11] A second set of questions, however, exposed a flaw in the manifesto—my focus on one aspect of aesthetic experience, the beautiful. Since I had written so extensively about postmodern conceptions of the postindustrial landscape sublime, I was blind to how narrow the manifesto appeared to those who did not know my earlier writing.[12] The following passages therefore elaborate on these aspects of the manifesto and offer more nuanced accounts of the entanglement of everyday life and aesthetic experience, socio-ecological ethos, and the often unexamined values embedded within the designed landscape.

ON BEAUTY

Beauty is a malleable category, and not a universal type. Contemporary designers create new forms of beauty. They stretch known categories through their exploration of new relationships and attempt to incorporate new functions and processes in their projects. I am impatient with designers who say that "the public" expects a particular kind of natural landscape beauty: this implies that the beautiful is only about the pleasurable and insipid, without regard for the sensual, the practical, and other emotions, or that beauty requires disinterestedness, a separation from the world, as if Kantian beauty was the only conception of beauty.[13]

Throughout history, we can find numerous examples of changing values about beauty in many practices and places. Didn't Frederick Law Olmsted and John Charles Olmsted create a public park in the 1880s predicated on an unfamiliar landscape type and experience, a saltwater marsh? Three decades ago, Anne Whiston Spirn wrote about their nervousness and audacity in doing so: "Olmsted felt the juxtaposition of the salt marsh and the city 'would be novel, certainly, in labored urban grounds, and there may be a momentary question of its dignity and appropriateness . . . but [it] is a direct development of the original conditions of the locality in adaptation to the needs of a dense community'" (Spirn 1984, 148, citing Zaitzevsky 1982, 57).

In another example, architect Ignasi Sola-Morales described the role of contemporary landscape photography in altering his appreciation of the urban landscape. He credited contemporary photographers' depictions of "forgotten" highway interchanges and abandoned postindustrial sites—depictions of spaces beyond conventional streets, parks, plazas, and gardens—for revealing the latent possibilities in *terrain vague*, the strange, interstitial spaces formed by the processes of modernization that lay outside the bounds of accepted urban design theory (Sola-Morales 1995, 120).

Twentieth-century art is also rife with accounts of the stretching of the boundaries of beauty.[14] Landscape theorist and historian John Beardsley suggested that surrealism, particularly Louis Aragon's fascination with "the marvelous" and André Breton's declaration that "Beauty will be convulsive," intersected neatly with my thinking about beauty's cultural and historical malleability. The key concepts of surrealism are grounded in Aragon's belief that the marvelous is found in the contradictions of the real. Breton called these concepts the veiled erotic, the expiration of movement, and the magical circumstantial. They resonate with counter-conceptions of beauty found in the works of Dirt Studio, Turenscape, Gilles Clément, and Peter Latz, to name a few designers who inspired my manifesto.

Re-reading Rosalind Krauss's essay on surrealism afforded me new tools for discussing beauty as an experience of excitement caused by *frisson*, dissonance, tension, delayed or disturbing pleasure (Krauss 1985/1991, 112–13). Designed landscapes located within forests disturbed by timbering, abandoned steel mills, and military installations, or mining sites degraded by toxic wastes, do not intrinsically contain new conceptions of beauty. However, through the interventions of a designer, new configurations and calibrations of landscape forms and processes come into being. What is perceived to be outside the bounds of landscape beauty is stretched. Thus new conceptions of landscape beauty can be convulsive, disturbing, and challenging; through them we confront the entanglement of personal consumption, waste, and the postindustrial site.

As an expression of cultural values and an extension of personal experience, beauty is malleable; hence the qualifications created by theorists and philosophers, such as difficult beauty, or convulsive beauty, or functional beauty (Parsons and Carlson 2008). Dissonance can also be an aesthetic experience. As Theodor Adorno wrote, beauty is not just about pleasure. It can originate in the ugly, its antinomy. "The definition of aesthetics as the theory of the beautiful is so unfruitful because the formal character of the concept of beauty is inadequate to the full content [*Inhalt*] of the aesthetic" (Adorno 1970, 50). Writers Adorno, Breton, and Aragon, as well as designers Julie Bargmann, Richard Haag, Peter Latz, Ken Smith, Ignasi Sola-Morales, and Kongjian Yu offer examples in words, images, and places of beauty. The beauty of their work, however, may be stretched beyond what is immediately and visibly pleasing to new forms of beauty that are, in Dave Hickey's

words, "strangely familiar," and that call into question assumptions about our perceptions of found and constructed nature, as well as our conceptions of sustainable living and making (Hickey 1993).

This explication of the ways that beauty can be stretched in response to changing functions or widened capacities for experiencing pleasure answers some of the skeptics who assumed I was calling for a return to conventional ideas of natural beauty, or a new formalism. But my failure to state more forcefully and convincingly that I was interested in sustainable aesthetics, not just sustainable beauty, was a fundamental flaw. And although I wrote with my prior writings on aesthetics, environmental perception, and ethics in mind, I did not mention them. In those writings I explored the postmodern sublime in two public parks built on postindustrial sites (Meyer 1998) as well as the influence of phenomenology, earth art, and site-specificity on the practice of recent landscape architecture in the United States (Meyer 2001) and the importance of scale and duration in the immersive experience of large parks built on disturbed sites (Meyer 2007). Without knowledge of my prior writings, especially my interest in how the eighteenth and nineteenth centuries stretched the beautiful and the sublime into strange new forms and experiences of the contemporary designed landscape, it could have been easy for readers to assume that I was only interested in beauty, one component of aesthetic experience.

Here, let me underscore that beauty is only one of many aesthetic experiences possible in the world. As my colleague Julie Bargmann says, "Let's name and claim" the others. In the same way that eighteenth-century art and landscape theorists created the concept of the picturesque to describe a new aesthetic experience between the beautiful and the sublime, we need to be attuned to new aesthetic experiences, to name them, to understand their agency, and to explore their possibilities in the designed landscape (Reimer 2010).

AESTHETICS: THE ART AND SCIENCE OF SENSE PERCEPTION AND COGNITION

In the manifesto, I defined aesthetics as the art and science of sense perception and cognition. Drawing on the writings of pragmatist and phenomenological philosophers as well as landscape architects, I understand aesthetics as an experience, not a formal language, not surface appearance (Berleant 1991; Dewey 1934; Spirn 1998). Aesthetic perception requires an exchange, a perceptual entanglement between a sensing body and the world; it requires a pause and duration. As an art and science of sensory perception and cognition, aesthetics is a

mode of interacting with and knowing the world. Aesthetic knowledge combines theory with method to guide action.

Like Spirn and Dewey, geographer Nigel Thrift grounds aesthetic experience in everyday life. He writes, "It is crucial to note here that aesthetics is understood as a fundamental element of human life and not just an additional luxury, a frivolous add-on when times are good. Postrel puts it thus: 'Aesthetics is the way we communicate through the senses. It is the art of creating reactions without words, through the look and feel of people, places and things.'" Thrift continues, "It is an affective force that is active, intelligible, and has genuine efficacy: it is both moved and moving. . . . It is a force that generates sensory and emotional gratification. It is a force that produces shared capacity and commonality. It is a force that, though cross-cut by all kinds of impulses, has its own intrinsic value" (Thrift 2010, 291–92). In landscape terms, aesthetics is not a synonym for that which is seen, or the surface appearance of things. Aesthetic experience occurs within an affective world that implicates bodies, objects, spaces, values, experiences, and networks. A theory and practice of landscape affects and effects would recognize that encounters between people and places are exchanges of emotions, agency, and energies.

By situating the design of landscapes within this strain of everyday aesthetic theory, and not limiting oneself to a Kantian state of disinterested contemplation, links can be drawn between design, ethos, politics, socio-ecological networks, and bodies. Cultural theorist Terry Eagleton implicates the aesthetic as follows: "That territory [between aesthetics and life] is nothing less than the whole of our sensate life together— the business of affections and aversions, of how the world strikes the body on its sensory surfaces, of that which takes root in the gaze and the guts and all that arises from our most banal, biological insertion into the world" (1990, 13). A landscape architecture inspired by Eagleton and Thrift's rich description of the aesthetic would do more than accommodate programs or afford immersive sensory experiences. It would calibrate the density, proximity, accessibility, and distribution of landscape spaces and surfaces relative to the desired effect of that strike on the individual and collective body. Those affective strikes would occur during the socio-ecological practices of everyday life, such as using (bathing to gardening), storing (from cisterns to rain gardens), and conserving (from native plants to grey water systems) water, in and through a designed landscape. Maria Hellstrom Reimer's concept of "Unsettling Eco-scapes" (2010) resonates with this practice of creating challenging, uncanny, at times discordant, everyday landscapes.

AESTHETIC EXPERIENCE REQUIRES DURATION

For designers of landscapes familiar with this literature, aesthetic experience is a slow process. It is not immediate, nor is it exhausted by a glance. It is not synonymous with the visual although it includes looking. These insights draw on philosophers spanning a century, from Henri Bergson to Arthur Danto and Alexander Nehamas, as well as contemporary art historians and neuroscientists exploring empathetic aesthetic responses through brain scans. This literature has rendered several of the following principles upon which "Sustaining Beauty" is based and through which I continue to refine my beliefs about the socio-ecological agency of the designed landscape:

1. Beauty is connected to appearance but not exhausted by it. Time is required to apprehend the beautiful (Nehamas 2007, 42).
2. Aesthetic experience is delayed, requires duration, and exists in the exchange between what ones sees/experiences and what one knows (Danto 1999; Nehamas 2007).
3. Beauty draws us near, and makes us want to know more and to act. It urges us to

create (Haidt 2006; Nehamas 2007, 76, 132; Scarry 1999).

4. Aesthetic experience builds a mode of intuition that combines feelings and knowledge. It produces its own form of cognition (Bergson 1907).[15]

Recent findings in neuroscience resonate with these musings. Art historian David Freedberg and neuroscientist Vittorio Gallese's recent experiments demonstrate that aesthetic responses are registered in the brain and manifested in somatic responses in the body. They found that visible traces of the artist's hand or mark, and vigorous handling of the painterly or sculptural medium, led to aesthetic response that was more than emotional. It was embodied and created empathy between the creator, the work, and the viewer (Freedberg and Gallese 2011). This begs the question whether perception of the conscious marks of the designer or gardener of landscapes—versus the recreation of natural beauty or the appearance of a wild ecosystem—might intensify affective bonds between humans and the nonhuman natural world.

Affects and Social Aesthetics

Why does a broader understanding of aesthetics matter to a sustainability agenda?

When talking about a delayed or prolonged experience in a public space, such as a park, a city street, or a campus, one cannot help but consider the collective experience of new configurations of urban nature. Can the experience on the ground of new socio-ecological entanglements—expanses of flooded or moist surfaces with novel material conditions or spatial practices—over time, result in more than self-absorbed reverie? Having wondered and worried about this when writing "The Post–Earth Day Conundrum" (Meyer 2001), I now realize that the duration of experience that I was considering through an aesthetic lens still needs to be broadened. We need to consider how landscape architects, in light of Guattari's "ethico-aesthetics," can "develop a creative responsibility for new modes of living as they come into being" (Bertelsen and Murphie 2010, 14). We need to explore how designing for the practice of everyday life in the designed landscape, an extended durational experience, can contribute to a new social aesthetics, a new ethos of sustainable perception and living.

Ben Highmore, a cultural theorist and scholar of everyday life, builds upon Gregory Bateson's 1958 concept of ethos as a *"culturally standardized system of organization of the instincts and emotions of the individuals,"* having "a definite set of sentiments towards the rest of the world," and "an emotional background" (2010, 128). In his persuasive argument for recasting aesthetics as a social issue, not a personal experience, Highmore describes the connection between ethos and social aesthetics in ways that resonate with landscape thinking and making: "Ethos, to borrow a term from Jacques Rancière, could be thought of as the 'distribution of the sensible' (*le partage du sensible*): 'the system of *a priori* forms determining what presents itself to sense experience. It is a delimitation of spaces and times, of the visible and invisible, of speech and noise, that simultaneously determines the place and the stakes of politics as a form of experience' (Rancière 2004, 13). Ethos, then, would be the orchestration of perception, sensorial culture, affective intensities, and so on: more pertinently it will be the interlacing of these" (Highmore 2010, 128–29).

This connection between landscape space, form, material, perception, and social aesthetics is just below the surface of what I have been writing and thinking about over the past decade. Several colleagues and students suggested that "theories of affect" might afford insights into specific issues that had troubled me since writing "The Post–Earth Day Conundrum" (Meyer 2001). Much writing about aesthetics is about individual experience; yet, numerous, simultaneous individual experiences in a public space

comprise an aesthetic collectivity and create new ways of living in and thinking about the environment. Thus, to better understand the agency of aesthetics in sustainability discourse, a promising area of research and speculation resides in what Patricia Clough describes as "the affective turn" (Clough 2007).

Affects impress on the emotions; they touch us and move our hearts. According to Baruch Spinoza, the seventeenth-century philosopher, affects connect the mind and body; they can be actions or passions. But unlike the interest in the body in either feminist theory or phenomenology, for instance, contemporary theories of affects entangle bodies with the world of technologies and networks, from the ecological to the logistical to information, suspending them in the space of duration during which time effects take place and affects unfold.[16]

This passage from Gregory Seigworth and Melissa Gregg's introduction to *The Affect Theory Reader* (2010) conveys the import for sustainable designers interested in the interconnections across scales and systems, from the body to the territory, from the biological to the technological: "Affect arises in the midst of *in-between-ness:* in the capacities to act and to be acted upon. . . . That is, affect is found in those intensities that pass body to body (human, non-human, part-body and otherwise), in those resonances that circulate about, between and sometimes stick to bodies and worlds. . . . Affect, at its most anthropomorphic, is the name we give those forces—visceral forces beneath, alongside, or generally *other than* conscious knowing, vital forces insisting beyond emotion—that can serve to drive us toward movement, toward thought and extension" (Seigworth and Gregg 2010, 1).

In the mid-1990s, a group of cultural critics and theorists including Brian Massumi, Gilles Deleuze and Felix Guattari, Eve Sedgwick and Adam Frank, developed strains of this particular approach to philosophy and cultural theory. They were reacting partly to the overly textual basis of structuralism and post-structuralism, and partly to the extraction of the phenomenological body from the entangled force field of flows, space, materials, and affects that constitute the world. However, aesthetic experience of landscape is inextricable from the world's complex, intertwining systems and flows. This may be the single most important trajectory that "Sustaining Beauty" has launched me on. It constructs a discursive field between writings on sustainability and aesthetics with those on landscape infrastructure and logistic landscapes. In doing so, it has the capacity to transform a landscape designed for individual sensation or experience to a landscape that reinforces a sensorial culture aware of how interlaced human routines are with the movement, as well as the scarcity or abundance, of materials and energies. Over the past few years, I have seen or read about several projects that deploy on-site affective intensities as a bridge to a broader network awareness, but Taylor Cullity Lethlean's new garden for Australia's Royal Botanical Garden was the first that I experienced and thought about through this lens.

Descriptive Entanglement: The Australian Garden at Cranbourne, Victoria

While on a "Sustaining Beauty" lecture tour in Australia, I visited a new botanical garden with jarring, unsettling, and saturated spatial sequences that alluded to national water-practices and gardening preferences. The institution's mission to propagate, display, and advocate for the native species and habitats of Australia maps readily onto this shift from a phenomenology of personal emotions to social affects and social aesthetics. The Australian Garden demonstrates several ways that encounters with the landscape may change bodies, from the individual to the collective, from the human to the nonhuman (such as an underappreciated plant species). The passage below describes my aesthetic experience, and allows others to witness, vicariously, how the

Fig. 2.1. The central Red Sand Garden, designed by Taylor Cullity Lethlean landscape architects in collaboration with sculptors Mark Stoner and Edwina Kearney. The 4-acre space is marked with drifts of thin, white, shiny striations and shallow dark-green vegetated disks. A crisp constructed horizon edits out most of the demonstration and exhibition gardens that cluster around the inaccessible center. Photograph by author.

garden's effects and affects were extended, prolonged, delayed, and entangled with the following two weeks of travel throughout Australia, and beyond. I hope my account of one landscape encounter allows others to see the potential of affect theory for creating, perceiving, and interpreting new modalities of sustainable landscape design. I offer up this garden encounter as an example of an imaginative counter-tradition of sustainable socio-aesthetics that values the fluctuating, the particular, and the dense force field that weaves together encounter and memory, event and network, perception and cognition, experience and action.

As I reach the bottom stairs, a shallow red-orange bowl fills my visual field. I have read that this inaccessible, tilted landform is a representation of the Australian continent—its dry, red sandy center of dune fields and ephemeral stream beds, and a sculpture designed by a landscape architecture firm and two sculptors. Colleagues in Melbourne had explained the meaning of the

Australian Garden's central space. Surface, lines and events—rendered in sand, ceramic and vegetation—evoke the yet unseen continental vastness [fig. 2.1].

But as soon as I turn my head and move my feet these descriptions are insufficient, replaced by a myriad of competing associations. The wide panoramic view that introduces a visitor to the Australian Garden is larger than one's cone of vision. Unable to take in the figured-ground plane in a glance, I turn to the left, where the red-orange sand garden slips below low billows of textural plantings, the edges obscure. What kind of sculpture folds this seamlessly into its setting? Without familiar plants to orient me and confronted by this odd, inaccessible, and assertive foreground, it is as if I am listening to a foreign language, trying to discern meanings from the context when the individual sounds do not congeal into a recognizable verbal pattern. If not for the gardeners—on their knees caring, weeding, and planting like the figures in a Bruegel engraving—I could easily forget I was in a botanical garden [fig. 2.2].

Fig. 2.2. The perimeter space to the left includes a sequence of demonstration gardens of various plant ecologies and habitats where gardeners speed up and slow down time through practices of care, management, and cultivation. The visitor must deploy a double gaze apprehending both the stable visual symbol and dynamic material edge. Photograph by author.

Fig. 2.3. The perimeter space to the right—known as the rockpool waterway—is visually and materially grafted into a notch in the sand garden, inserted into a joint of dry sand. The location and shape of that graft—and the abrupt shifts of scale, material, and phenomena that are contained in this constructed edge—elicit vivid reactions. Photograph by author.

In their cultivation of the plants, these gardeners remind me of the purpose of this place. This is not a sculpture garden, or a garden with a sculpture in the middle of it. The actions of the gardeners, on the margin between the surreal otherness of the red-orange sand garden and the public path, make visible the propensity of this landscape. It exists between a visual symbol of another landscape, transposed from the continent's interior, and an embodied, dynamic, unfolding landscape.[17]

I then look to the right, passing my eyes along the folded, dune-like edge of the red-orange center as it slopes down towards a thin green verge and a horizontal channel full of water and reflected sunlight [fig. 2.3]. My initial reading of the center as a symbol, or a sculpture, or perhaps a scenic view but not experience, can no longer be sustained. I am jolted out of my comfortable assumption that, after all the shock of the entrance, this is a typical botanical garden of tended plots. Adjacent to this barren, inaccessible, high and dry, sculptural center is a beckoning space of extreme wetness. These conditions are juxtaposed so abruptly and the state shift from dry to wet so compressed, I struggle for orientation. My feet are compelled towards this spatial and material joint, this condition of oscillation between sand and water, dry and wet. I am engaged, immersed, and curious.

Botanical gardens are institutions for research and education as well as entertainment. Their missions have moved beyond nineteenth-century aspirations to collect, catalog, and curate. The Australian Garden's sponsoring organization, the Royal Botanic Garden, defines itself as a leader in "the conservation of plants . . . through biodiversity research, conservation programs to protect rare and endangered species and the study of habitats" (Roberts 2007, inside cover). Clearly, Taylor Cullity Lethlean's and their consultant Paul Thompson's knowledge of native plants, as well as their arrangement of the beds surrounding the sand garden, fulfill this mission. The gardens bring the public into close encounter with native plants that may be unfamiliar to them.

But, how does this dissonant entrance into the botanical garden contribute to that mission? From one's first encounter in the garden, plants are understood in relationship to site and habitat. Yes, the sand garden can be understood as a synecdoche, or fragment, that stands for the relationship between the continent's dry interior and its more temperate, vegetated coastal perimeter. But the patterns and arrangement of plants are also related to the availability of moisture, and to processes of land formation and water flows here in Cranbourne. The garden quickly shifts in one's mind from a miniature and an abstraction to an actual landscape

comprised of varied "to-scale" habitats, microclimates, and related plant groupings. One's experiential apprehension of plants in specific situations and habitats in the garden increases one's familiarity with plants and, more generally, the conservation of plants in their native habitats (fig. 2.4). As Seigworth and Gregg note, a body is never alone; it is

Fig. 2.4. A series of native-plant habitat gardens flank the western edge of the sand garden. The red-orange sand surface appears and disappears as one walks along the perimeter path and through the gardens. Photograph by author.

Fig. 2.5. The original Royal Botanical Garden in South Yarra, Melbourne, created in the nineteenth century. Its picturesque gardens of subtropical plants drew on an established conception of landscape beauty replete with lawns that were familiar to British colonists of the time. Photograph by author.

"always aided and abetted by, and dovetails with, the field or context of its force-relations" (2010, 3).

How is this achieved in design terms? Lethlean spent a year studying gardens in Japan when he was in college. His account of the lessons gleaned from observing and documenting these gardens, many of them *kare sansui*—dry sand or rock gardens—expanded my interpretation of the Australian Garden. So did his mention of Fred Williams's abstract paintings of the Australian landscape that have inspired and influenced the Taylor Cullity Lethlean design team.[18] I have imaginatively revisited the garden numerous times—studying the photographs I took on site and recalling the sequence they record. My newfound knowledge of these multivalent references— the disparate force fields of geomorphology, native habitats, abstract modern art, and traditional Japanese garden theory—filtered through the eyes of an Australian-born and -educated landscape architect, imbues the garden with an even stranger beauty than before.[19]

The Cranbourne garden engaged my feet, my eyes, my body, and my mind over the following weeks and months. During the time in the garden, my awareness of the Australian landscape began. As I traveled, I returned to the garden, mentally, re-calibrating my reactions. The garden's plants and materials,

Fig. 2.6. The vast dry landscape of Australia's continental interior viewed from an airplane, 2010. Photograph by author.

and their arrangements, swirled within a field of flows and forces as large as the continent, as old as the English-picturesque botanical garden in Melbourne and the geological morphology I studied from the plane (figs. 2.5 and 2.6).

I now appreciate the connection between places and affects as being less about one-to-one correlation to morality, ethics, and care—issues I explored in previous writings— and more about opening up the possibility of altering how people relate to one another and the natural world. This "maybe," or promise, of affect is captured by Lauren Berleant in her modest proposition that "the substitution of habituated indifference with a spreading pleasure might open up a wedge into an alternative ethics of living, or not" (2010, 105). Seigworth and Gregg suggest, "Maybe that is the 'for-now' promise of affect theory's 'not yet,' its habitually rhythmic (or near rhythmic) undertaking: endeavoring to locate that propitious moment when the stretching of (or tiniest tear in) bloom-space could precipitate something more than incremental. If only. Affect as promise" (2010, 12). A landscape architecture of affects cannot be substantiated through metrics known in advance; it cannot guarantee changes in what people think about the environment, or do in the landscape. A landscape architecture of affects, and the experiments in living that occur as the public adapts to, and adapts, the designed landscape will be provisional and speculative. What if?

After visiting designed landscapes such as the Australia Garden, I believe new sustainable landscape aesthetics and practices might best emerge from what Highmore calls "experiments in living" (2010, 135) and making, from explorations of the lived experience of place created within a network of enigmatic materials, strange objects, and multiple bodies interacting and pulsating over time. We need more slow landscapes, wherein aesthetic experience occurs over time and is shaped by a new "materialism where a body would be understood as a nexus of finely interlaced force fields" (Highmore 2010, 119). We need to find ways to explore and document the affects and entanglements, the "dense weave of aesthetic propensities" that occur in designed encounters over time (Highmore 2010, 135). This might require

ways of considering, through the lens of neuroscience, how the sight of gardeners tending a landscape throughout the seasons might prompt a somatic response in the visitor and alter her appreciation for the bonds of care, manifest in the movements of a shoulder, an arm, and a shovel or rake, that connect the human and nonhuman world. Or it might require us to develop different genres of landscape criticism that more intricately enmesh a designed landscape in its milieu, or logistics landscape.[20]

This revelation, that aesthetic intuition is a form of cognition connecting bodies, experiences, feelings, places, networks, and actions, intrigues me from a sustainable-design perspective. Many authors have identified the gap between what we know about large systems and long processes, and what we actually do in our daily lives, as one of the biggest obstacles to sustainable life changes (Gertner 2009). Recently, I heard that Taylor Cullity Lethlean has designed prototypical vernacular residential gardens based on the native species and planted form experiments they undertook at Cranbourne.[21] I am intrigued with the complex web of connections that center on this botanical garden, from the scale of the residential garden to the continent's geomorphology, from the rockpool waterway to the region's scarce rainwater, from Williams's stroke of paint on a canvas to a gardener's cultivation of a native plant, formerly considered a weed.

Aesthetics, Affects, Sustainability, and Public Policy

It might be easy to discount my explorations as academic speculations that are not urgent in this period of economic retrenchment and living within our means. Instead, one might argue, we should focus on landscape infrastructure because it is necessary and may be less expensive than traditional centralized engineering approaches—except for one thing. The 2005 United Nations' Millennium Ecosystem Assessment described aesthetics as one of the four types of ecosystem services that support human life: provisioning services, regulating services, supporting services, and cultural services, such as aesthetic, recreational, spiritual (Millennium Ecosystem Assessment 2005). While this august group of scientists from around the world included aesthetic experience as a critical function of ecosystems, there has been little research published in the last decade on how to evaluate the cultural services provided by ecosystems. Some recent scholars have suggested that the category of cultural services should be jettisoned because there are not enough metrics to substantiate their inclusion. Others have suggested that cultural

services should be downgraded to benefits, not services, for the same reason (Boyd and Banzhaf 2007; Kramer 2007; Layke 2009).

In what has been written, there is little recognition that there are several kinds of ecosystems to consider—from the remote and relatively undisturbed to the made, the novel, the urban, and the designed—and that their aesthetic effects and affects might vary. I hope that some readers will agree that the answer is more research, now. I trust that other landscape architects agree that the aesthetic affects of constructed and found ecologies are different. Surely, constructed ecosystems that function as public spaces perform differently aesthetically and create affects different from those found in the wild, in a nature reserve, or a national park (Nassauer 1997; Parsons and Carlson 2008).

Shouldn't we argue that the aesthetics of constructed ecosystems are ecosystem services because of the effects and affects they may have on the communities that live near them and frequent them? In *Design for Ecological Democracy*, Randy Hester made a similar argument with his concept of "Impelling Form" (2006). These are not just individual emotional or psychological affects in an Olmstedian mode. They are collective affects that simultaneously tap into "already existing structures of feeling" but intensify them through prolonged, vivid, and strange encounters with constructed nature.

These affects suspend us between here (event, place) and there (watershed, territory, logistics landscape), between the place and the network. Bateson and Highmore suggest that ethos, the "dense weave of aesthetic propensities shared by a group" or social aesthetics, can be changed through "experiments in living" (Highmore 2010, 135). Kate Soper, the philosopher, urges us to seek the "already existing structures of feeling" that are changing how everyday life is practiced and how everyday space is conceived (2008, 576). Felix Guattari, the philosopher, calls for a political approach to aesthetics, calling for the development of a "creative responsibility for modes of being as they come into being" (Bertelsen and Murphie 2010, 141). From their varied points of view, these scholars suggest that changes in practices are felt and lived into being through the life and work of sentient beings.

Landscape architecture is the discipline that designs the spaces where these conditions and relationships are perceived, studied, and celebrated by scholars in anthropology, philosophy, sociology, and the sciences. We are designing these "experiments in living" in the entanglement of social practices and living systems. We need to be reflective about them, claim them, and try to measure their qualitative contributions to a sustainable future. This will require us to find bridges between existing scholarly pursuits—

between the body, landscape tectonics/materials research, ecological landscape performance standards, and urban design. We can extract some of the sensibilities from affect theory, but must invent ways to ground this intellectual position in the design process, in more complex understandings, and representations of spatial practices and program, and in our post-construction evaluation of projects. If we do, new forms and practices of social aesthetics might emerge from these endeavors.[22]

As I write this, I am more modest in my hopes for sustaining beauties, aesthetic performance, and landscape architecture than I was in 2007. At the very least, I am less ardent, and content with exploring the connections between aesthetics and ethos, instead of ethics.[23] I now consider aesthetics as an entanglement that implicates the social and embraces the drive to movement as a promise, but not necessarily an outcome. Social aesthetics and ethos are unhinged from morality and certainty. Perhaps this reflects the broader zeitgeist, political as well as professional. Regardless, this qualified optimism is a quality that serves designers well. Through our words, images, and built works, we set the world in motion. We chart out propensities without controlling outcomes. We design socio-ecological experiments in living with no promises. Something. Perhaps. It might.

Who knows? Insisting on the imperative to experiment and to work without certainty, in the face of increasing pressure to adopt best management practices and to establish ecosystem service metrics, may be the most pressing challenge for the next generation of landscape architects who care to design places that matter culturally as well as ecologically.

NOTES

1. "Sustaining Beauty: The Performance of Appearance" originated as a lecture for two different audiences outside the United States. Immediately after my Beijing lecture ("Landscape Architecture for Ecological Security," Peking University, September 22–23, 2007), the *Journal of Landscape Architecture* (JOLA) published the manifesto in its raw form, largely as delivered, with a few footnotes added (Meyer 2008a).

2. Delivering the manifesto opened up many avenues of research. I came to appreciate the pervasive critiques of sustainability discourse that existed in other disciplines—across the humanities and sciences. I confess I was naive about the debates, especially given the singular and limited discussion of sustainability within the design fields. I learned of some of these critical positions at the second 2007 venue for "Sustaining Beauty," the Royal Geographical Society annual meeting in London. Upon returning from that event, I immersed myself in the sustainability discourse in the humanities and sciences with the assistance of UVA graduate students in my "Situating Sustainability" seminars, several of whom developed thesis projects on related topics. These UVA students, now alumni, include Kate Boles, Aja Bulla-

Richards, Alexa Bush, Toshi Karato, Shanti Levy, Jen Lynch, David Malda, Andrea Parker, Julia Price, and Kurt Petschke.

3. Editor William Thompson asked if "Sustaining Beauty" could be republished in *Landscape Architecture Magazine*. That issue was distributed at the 2008 ASLA meeting in Philadelphia, where I also participated in a session on landscape architecture, metrics for sustainable aesthetics, and ecosystem services.

4. The current essay is an intellectual and political work in progress, some of which was developed in a shorter piece, "Slow Landscapes: A New Erotics of Sustainability" (2009–10), some for a book on Taylor Cullity Lethlean, an Australian landscape architecture firm (Lee and Ware 2014), and some of which is still brewing. For younger scholars, I offer these musings to underscore the importance of writing about topics that matter to nonacademics, and of finding links between academic discourse and the profession.

5. For more on this issue, see Meyer (1998), "The Expanded Field of Landscape Architecture."

6. See interview with Michael Van Valkenburgh and Matt Urbanski about their concept, "Hypernature" (Amidon 2005), 56–72.

7. My position here is based in Olmsted's 1868 writing about landscape aesthetics and role of public parks: "A park is a work of art designed to produce certain effects upon the mind of men" (147–57), and on philosopher Elaine Scarry's 1999 account of the power of beauty to de-center ourselves, to prompt us to want, and to care.

8. For divergent positions on the relationship between place, process, fitness, and beauty, see Eaton (1997) and Berrizbeitia (2007).

9. The authors I drew upon for these tenets, such as John Dewey (1934), Anne Whiston Spirn (1988), and Elaine Scarry (1999), focus on individual aesthetic and environmental experiences, but not on social aesthetics. So, although my prior work

speculated upon the broader, collective impact of many environmental/aesthetic experiences, I did not yet have a framework for social aesthetics, such as theories of affect.

10. See Carlson and Lintott's anthology (2008) for several excellent articles on natural beauty, natural aesthetics, and its distinction from artistic beauty.

11. This was introduced to me by Peter Connolly, an Australian colleague now teaching in New Zealand.

12. Maria Hellstrom Reimer's "Unsettling Eco-scapes" was the most insightful and influential of those responses.

13. Those unfamiliar with aesthetic theory may be interested in Mark Foster Gage, *Aesthetic Theory: Essential Texts for Architecture and Design* (2011) for his short critical essay and excerpts of twenty key texts, including Kant's *Critique of Judgment.*

14. John Beardsley, the art and landscape historian, reminded me of this in his extremely thoughtful response to hearing my lecture "Sustaining Beauty" at the 2008 European Landscape Biennale in Barcelona.

15. Gage introduces Bergson's 1903 writing as follows. Bergson "combines the accepted observation-based tactics of scientific rationalism with the possibility of additional information being gathered by less tangible aesthetic forms of perception. . . . For Bergson, intuition, an aesthetically tinged and disinterested refinement of raw human instinct, exists as the equally contributing counterpoint to raw scientific intelligence—both of which are required for one to possess true knowledge" (Bergson 1903, excerpt in Gage 2011, 154–55).

16. For this reason, I think that theories of affects could benefit nicely from the logistic landscape thinking and mapping of Pierre Belanger and Charles Waldheim, but that particular inquiry is outside the scope of this paper.

17. For an account of propensity, or *shi*, which "oscillates between the static and dynamic point of view" and can be found in "any given configuration" in

"the inherent propensity for the unfolding of events," see Fung (1999, 144–45).

18. A year after I visited the Australian Garden, Perry Lethlean lectured at the American Society of Landscape Architects' (ASLA) annual meeting in Washington, DC, and at the University of Virginia School of Architecture. I gained additional insights into the dense entanglement of Australian design aesthetics, environmental attitudes, and cultural conceptions of sustainability from these lectures and several informal conversations with Lethlean and others who know the garden.

19. This passage is from an essay I wrote for Gini Lee and Sue Anne Ware (2014).

20. Anita Berrizbeitia's 2005 book, *Roberto Burle Marx in Caracas,* is a rare example of a recent piece of landscape architectural history and criticism that covers this subject well.

21. Jen Lynch worked as a summer intern at Taylor Cullity Lethlean while a University of Virginia graduate landscape architecture student. She relayed this development to me upon her return to studies in fall 2011.

22. Jane Hutton's recent scholarship, presented at the CELA 2012 conference, and her 2013 article "Reciprocal Landscapes: Material Portraits in New York City and Beyond," is a promising examples of this type of intellectual enterprise.

23. My thanks to Dilip da Cunha for several conversations in Philadelphia and Melbourne about the unexamined moral implications of the manifesto. These allowed me to see the distinction between morality and ethos as one that is key to an open, creative exploration of the connection between environmentalism and aesthetics.

REFERENCES

Adorno, Theodor W. 1970. *Aesthetic Theory.* Trans. Robert Hullot-Kentor. Minneapolis: University of Minnesota Press, 1997.

Amidon, Jane. 2005. *Michael Van Valkenburgh Associates: Allegheny River Park.* New York: Princeton Architectural Press.

Benezra, Neal David, and Olga M. Viso, eds. 1999. *Regarding Beauty: A View of the Late Twentieth Century.* Washington, DC: Hirshhorn Museum.

Bergson, Henri. 1903 and 1907. "From *Creative Evolution*" (orig. pub. 1907) and "From *Introduction to Metaphysics*" (orig. pub. 1903). Excerpts reprinted in *Aesthetic Theory,* ed. Gage, 153–59.

Berleant, Arnold. 1991. *Art and Engagement.* Philadelphia: Temple University Press.

Berleant, Lauren. 2010. "Cruel Optimism." In *The Affect Theory Reader,* ed. Gregg and Seigworth, 93–117.

Berrizbeitia, Anita. 2005. *Roberto Burle Marx in Caracas: Parque del Este, 1956–61.* Philadelphia: University of Pennsylvania Press.

———. 2007. "Replacing Process." In *Large Parks,* ed. Julia Czerniak and George Hargreaves, 175–98. New York: Princeton Architectural Press.

Bertelsen, Lone, and Andrew Murphie. 2010. "An Ethics of Everyday Infinities and Powers: Felix Guattari on Affect and Refrain." In *The Affect Theory Reader,* ed. Gregg and Seigworth, 138–57.

Boyd, James, and Spencer Banzhaf. 2007. "What Are Ecosystem Services? The Need for Standardized Environmental Accounting Units." *Ecological Economics* 63 (2–3): 616–626.

Carlson, Allen, and Sheila Lintott, eds. 2008. *Nature, Aesthetics and Environmentalism: From Beauty to Duty.* New York: Columbia University Press.

Clough, Patricia Ticineto, with Jean Halley. 2007. *The Affective Turn: Theorizing the Social.* Durham, NC: Duke University Press.

Danto, Arthur. 1999. "Beauty for Ashes." In *Regarding Beauty: A View of the Late Twentieth Century,* ed. Neal Benezra and Olga M. Viso, 183–97. Washington, DC: Hirshhorn Museum Press.

Deleuze, Gilles, and Felix Guattari. 1987. *A Thousand Plateaus: Capitalism and Schizophrenia.* Trans.

Brian Massumi. Minneapolis: University of Minnesota Press.

Dewey, John. 1934. *Art and Experience.* New York: Minton, Balch & Co.

Eagleton, Terry. 1990. *The Ideology of the Aesthetic.* Cambridge, UK: Basil Blackwell.

Freedberg, David, and Vittorio Gallese. 2011. "Motion, Emotion, Empathy in Esthetic Experience." In *Aesthetic Theory,* ed. Gage, 309–24.

Fung, Stanislaus. 1999. "Mutuality and Cultures of Landscape Architecture." In *Recovering Landscape: Essays in Contemporary Landscape Architecture,* ed. James Corner, 140–151. Princeton, NJ: Princeton Architectural Press.

Gage, Mark Foster, ed. 2011. *Aesthetic Theory: Essential Texts for Architecture and Design.* New York: W. W. Norton & Co.

Gertner, Jon. 2009. "Why Isn't the Brain Green?" *New York Times Sunday Magazine: The Green Issue* (April 19): 36–43.

Gregg, Melissa, and Gregory J. Seigworth, eds. 2010. *The Affect Theory Reader.* Durham, NC: Duke University Press.

Haidt, Jonathan. 2006. *The Happiness Hypothesis.* New York: Basic Books.

Hester, Randolph T. 2006. *Design for Ecological Democracy.* Cambridge, MA: MIT Press.

Hickey, Dave. 1993. *The Invisible Dragon: Four Essays on Beauty.* Los Angeles: Art Issues Press.

Highmore, Ben. 2010. "Bitter after Taste: Affect, Food and Social Aesthetics." In *The Affect Theory Reader,* ed. Gregg and Seigworth, 118–37.

Hutton, Jane. 2013. "Reciprocal Landscapes: Material Portraits in New York City and Beyond." *Journal of Landscape Architecture* (Europe), vol. 8 (Spring): 40–47.

Kramer, Randall. 2007. "Economic Valuation of Ecosystem Services." In *The Sage Handbook of Environment and Society,* ed. Jules Pretty et al., 171–80. London: Sage.

Krauss, Rosalind. 1985/1991. "The Photographic Conditions of Surrealism." In *The Originality of the Avant-Garde and Other Modernist Myths*, 87–118. Cambridge, MA: MIT Press.

Layke, Christian. 2009. "Measuring Nature's Benefits: A Preliminary Roadmap for Improving Ecosystem Service Indicators." *WRI Working Paper* (September), 36 pp. Washington, DC: World Resources Institute.

Lee, Gini, and SueAnne Ware. 2014. *Making Sense of Landscape: Taylor Cullity Lethlean*. Easthampton, MA: Spacemaker Press and ORO Editions.

Massumi, Brian. 2002. *Parables for the Virtual: Movement, Affect, Sensation*. Durham, NC: Duke University Press.

Meyer, Elizabeth K. 1997. "Seized by Sublime Sentiments." In *Richard Haag: Bloedel Reserve and Gas Works Park*, ed. William Saunders, 5–28. New York: Princeton Architectural Press.

———. 1998. "The Expanded Field of Landscape Architecture." In *Ecological Design and Planning*, ed. George F. Thompson and Frederick R. Steiner, 35–79. New York: John Wiley & Sons.

———. 2001. "The Post–Earth Day Conundrum." In *Environmentalism in Landscape Architecture*, ed. Michel Conan, 187–244. Washington, DC: Dumbarton Oaks.

———. 2007. "Uncertain Parks: Citizens, Disturbed Sites and a Risk Society." In *Large Parks*, ed. Julia Czerniak and George Hargreaves, 58–85. New York: Princeton Architectural Press.

———. 2008a. "Sustaining Beauty: The Performance of Appearance." *Journal of Landscape Architecture* (Europe), vol. 3 (Spring): 6–23.

———. 2008b. "Sustaining Beauty: The Performance of Appearance." *Landscape Architecture Magazine* 98 (October): 92–131.

———. 2009–10. "Slow Landscapes: A New Erotics of Sustainability." *Harvard Design Magazine Special Issue on Sustainability + Pleasure* (Fall/Winter): 22–31.

———. 2013. "Grafting, Splicing, Hybridizing: The Strange Beauties of the Australia Garden." In *Taylor Cullity Lethlean: Making Sense of Landscape*, ed. Gini Lee and SueAnne Ware, 56–69. Easthampton, MA: Spacemaker Press and ORO Editions.

Millennium Ecosystem Assessment. 2005. *Ecosystems and Human Well-Being: Synthesis*. Washington, DC: Island Press.

Nassauer, Joan. 1997. "Cultural Sustainability: Aligning Aesthetics and Ecology." In *Placing Nature: Culture and Landscape Ecology*, ed. Joan Nassauer, 64–84. Washington, DC: Island Press.

Nehamas, Alexander. 2007. *Only a Promise of Happiness: The Place of Beauty in a World of Art*. Princeton, NJ: Princeton University Press.

Olmsted, Frederick Law. 1868. "Psychological Effect of Park Scenery: Address to (the) Prospect Park Scientific Association." In *The Papers of Frederick Law Olmsted: Supplemental Series*. Volume 1 of *Writings on Public Parks, Parkways, and Park Systems*, ed. Charles E. Beveridge and Carolyn F. Hoffmann, 147–57. Baltimore: Johns Hopkins University Press, 1997.

Parsons, Glenn, and Allen Carlson. 2008. *Functional Beauty*. Oxford, UK: Oxford University Press.

Rancière, Jacques. 2004. *The Politics of Aesthetics: The Distribution of the Sensible*. London: Continuum Press.

Reimer, Maria Hellstrom. 2010. "Unsettling Eco-scapes: Aesthetic Performances for Sustainable Futures." *Journal of Landscape Architecture* 9 (Spring): 24–37.

Roberts, Bev. 2007. *The Australian Garden: A Unique Garden for the 21st Century*. South Yarra, Victoria: Royal Botanic Gardens Board.

Scarry, Elaine. 1999. *On Beauty and Being Just*. Princeton, NJ: Princeton University Press.

Sedgwick, Eve Kosofsky. 2003. *Touching Feeling: Affect, Pedagogy, Performativity*. Durham, NC: Duke University Press.

Seigworth, Gregory J., and Melissa Gregg. 2010. "An Inventory of Shimmers." In *The Affect Theory Reader,* ed. Gregg and Seigworth, 1–25.

Sola-Morales, Ignasi. 1995. "Terrain Vague." In *ANY Place,* ed. Cynthia Davidson, 118–24. Cambridge, MA: MIT Press.

Soper, Kate. 2008. "Alternative Hedonism, Cultural Theory and the Role of Aesthetic Revisioning." *Cultural Studies* 22:5 (September): 567–87.

Spirn, Anne Whiston. 1984. *The Granite Garden. Urban Nature and Human Design.* New York: Basic Books.

——. 1988. "The Poetics of City and Nature: Towards a New Aesthetic for Urban Design." *Landscape Journal* 7:2 (Fall): 108–26.

——. 1998. *The Language of Landscape.* New Haven, CT: Yale University Press.

Thrift, Nigel. 2010. "Understanding the Material Practices of Glamour." In *The Affect Theory Reader,* ed. Gregg and Seigworth, 289–308.

Zaitzevsky, Cynthia F. 1982. *Frederick Law Olmsted and the Boston Park System.* Cambridge, MA: Harvard University Press.

KATHRYN MOORE

The Value of Values

A radical redefinition of the relationship between the senses and intelligence is way overdue. Although I am writing from the perspective of an experienced teacher and practitioner of landscape architecture, these issues are equally relevant to architecture, urban design, and other art and design disciplines, as well as philosophy, aesthetics, and education more generally.[1]

Context is important. Today, there is a moment of profound change in the way we value the material, social, and cultural contexts of our lives. The best contemporary projects in Europe and elsewhere around the world demonstrate that it is no longer enough to simply "consider" the landscape or "take it into account." The landscape is not an afterthought, the bits left in between the buildings, developments, highways and town centers, or a vague blanket cover or surface that will look after itself. Given its proper status, landscape is the context upon and within which all developmental processes take place, and with all of its cultural, social, and physical potential, the landscape, the *idea* of landscape, is seen as a base layer against which decisions about all future development need to be made.

To have any real chance of providing a sustainable and lasting blueprint for the landscape, a new way of working needs to become wholeheartedly absorbed into all of the decision-making institutions and organizations responsible for policy and

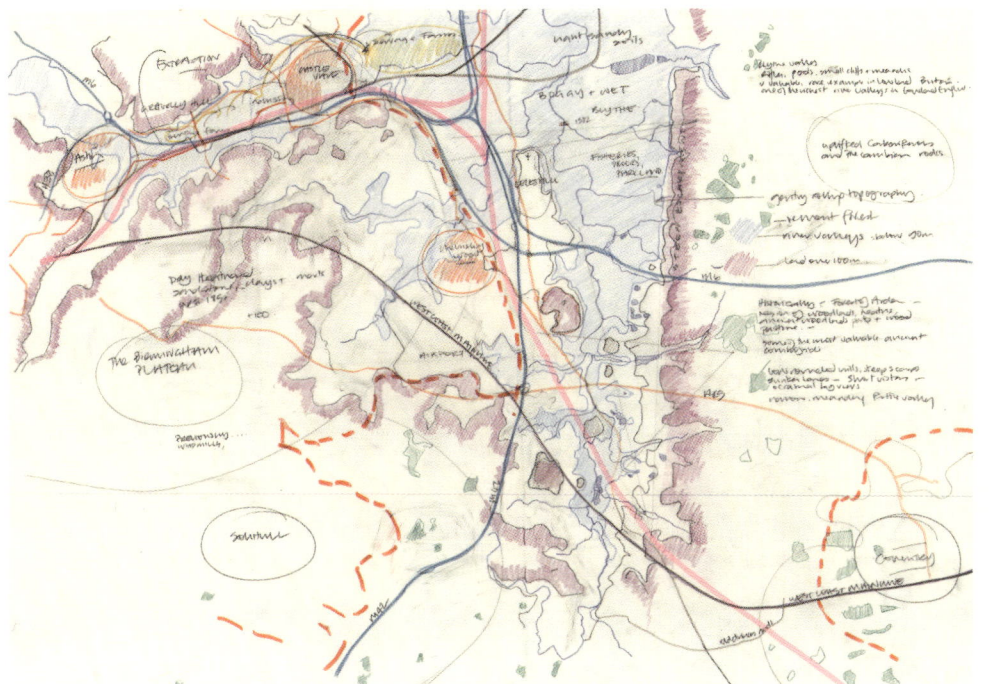

Fig. 3.1. High-Speed 2 Rail Link Landscape Vision (HS2LV) conceptual sketch and site analysis: tracing the remnants of the forest of Arden, the fisheries and ancient countryside of the boggy Blythe Valley, the heathlands, the track of quarries, sewage farms, and industry along the Tame. Designed and drawn by Kathryn Moore, 2012. Clients: Birmingham City Council, Solihull MBC, Birmingham Chamber of Commerce, Centro, and Arups. Used by permission.

strategic or regional planning at a national and international level, as well as, of course, education.

There is evidence of a sea-change in planning and development hierarchies with the landscape determined as the lead driver. This important economic and social concern is now firmly on the mainstream political agenda, and can be seen in projects such as HS Landscape Vision (HS2LV). This project proposes to transform the High-Speed 2 Rail Link, a highly contentious high-speed train line between London and Birmingham, into an iconic city-to-city national landscape infrastructure project that will play a significant role in shaping the United Kingdom's response to major environmental challenges (figs. 3.1 and 3.2). As a pilot study for the future, the study aims to put the Greater Birmingham region at the forefront of sustainable spatial development, conservation, and urban regeneration.[2]

Expanding the conceptual agenda and territorial scope of an engineering project, the proposal places the landscape at the core of HS2 and uses it as a catalyst for the economic, physical, and ecological transformation of communities affected by the route. Landscape architects are ideally placed to capitalize on this moment of change and take the lead in sustainable economic development—not just to deal with the details of technology and ecology, but also to address the bigger picture—informed

Fig. 3.2. HS2LV conceptual sketch. Bringing together nature and culture, art and science, economics and health, defining the city boundary, demonstrating a powerful and new symbiotic relationship between the towns, cities, and countryside, the drawings present a new identity for the region, an overarching vision to act as the context for future development. Designed and drawn by Kathryn Moore, 2012. Used by permission.

by a landscape-led, conceptually driven approach.

We have explored this approach in design studios at the Birmingham Institute of Art & Design (BIAD). Drawing together transport, health, economics, and culture, with a geographical sensibility and an acute awareness of the potential of the *idea* of landscape, final-year postgraduate students at BIAD create a city diagram. This work provides the context for further research and exploration and serves as the conceptual and artistic basis of their major design thesis. The students' city diagrams have sparked debate among public officials and stakeholders

in Washington, East Birmingham, and Kidderminster, not about vague intentions but about spatial ideas that raise significant questions about real situations.

Figures 3.3 and 3.4 are examples of city diagrams prepared by BIAD students for a study of Kidderminster. This kind of inquiry requires reevaluating many of the assumptions held about the nature of the visual, aesthetics, the design process, and landscape, planning, strategy, and policy. The first step towards reevaluation is in developing a pragmatic, holistic approach to landscape, the design process, and perception.

New model
settlement
within heathland
environment

Waterfront town
park and
Exposition Centre

New woodland
park and setting
for Exposition
exhibits replaces
ringroad

Comberton
Gateway and
transport hub

Production
and
marketplace
zone

Retrofit
residential and
business zone

Wetland Expo
accommodation
and nature
reserve

Fig. 3.3. City diagram (Kidderminster) produced for postgraduate diploma program, Birmingham Institute of Art & Design. The basis of this strategic vision for Kidderminster is a Europe-wide exposition themed around sustainable postindustrial regeneration, and set within the climate-change agenda. Using water as a metaphor for change, also to reflect both the reason for the town's location and the importance of the hydrological cycle in climate change, the city diagram explores the composition of the site and its component relationships. Designed and drawn by Michelle Anderson, 2011. Used by permission.

Fig. 3.4. City diagram (Kidderminster) produced for postgraduate diploma program, Birmingham Institute of Art & Design. The project proposes a new National Triathlon Centre and Recreational Parklands as venues for community and national events. The strategy overcomes the town's fractured geography, which includes broken nature corridors, underused waterways, and a dislocated sports and educational network. Designed and drawn by Tom Green, 2011. Used by permission.

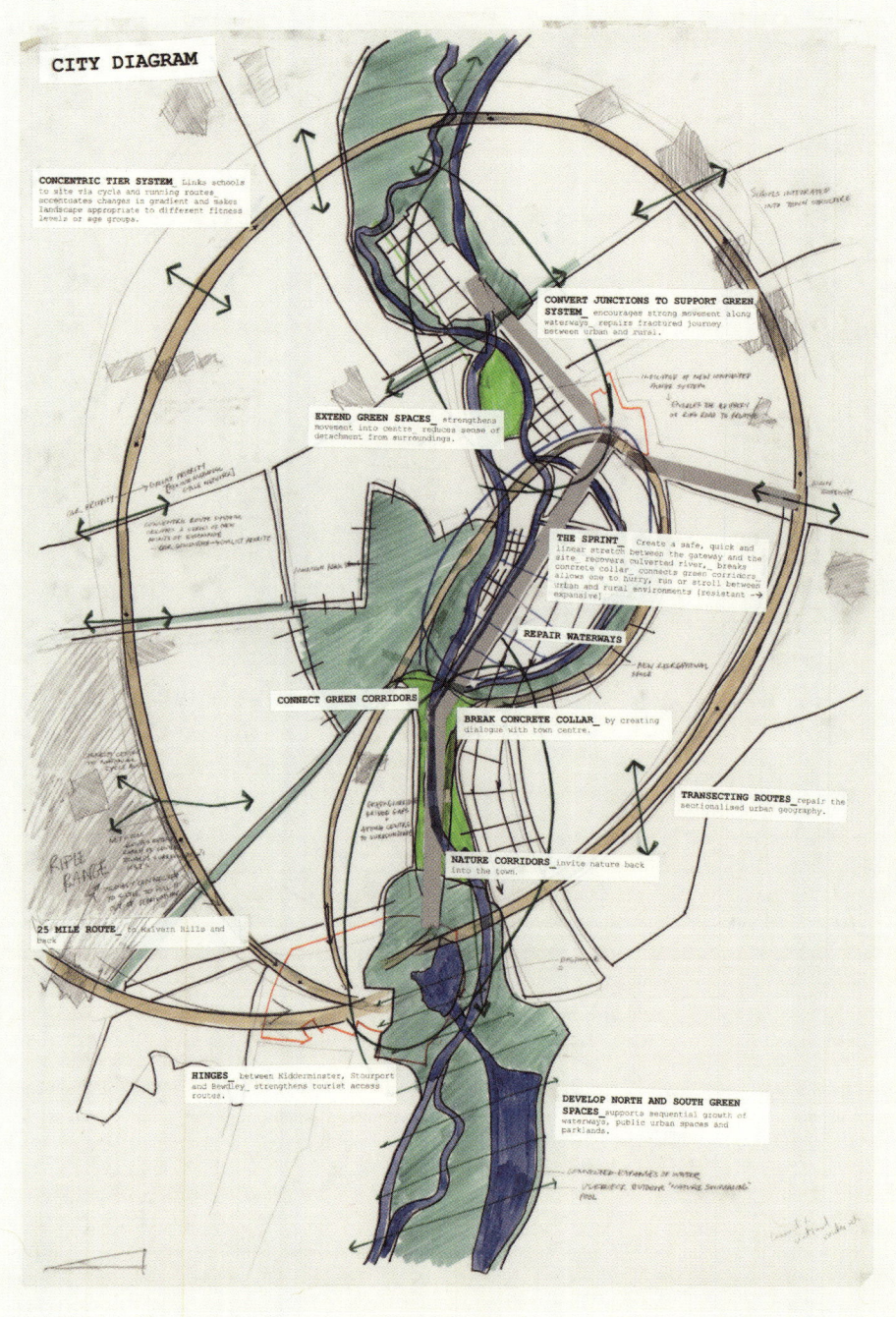

Design Dichotomy

In reevaluating the importance of landscape, the role of design is far from clear; in fact, it seems rather tenuous, fragile, and easily dismissed. It is, of course, possible to teach many aspects of design—design theory, criticism, history, technology, and modes of communication. But teaching the real nitty-gritty of the discipline, the designing part of design, is clouded for most by an air of subjectivity, and therefore is seen, somehow, as impossible to teach. Spectacularly ill-defined, design is often seen as a highly personal, mysterious act, almost like alchemy. And then there is the dangerous idea that it is possible, indeed preferable, to hide behind the supposed objective neutrality implied by the more "scientific," technology-based, problem-solving approaches. We even hesitate about defining what it is that designers do, considering talk about aesthetic and artistic sensibility, let alone design expertise, to be oxymoronic—a contradiction in terms. My work, therefore, is concerned with developing a greater understanding as to what designing means, in both theory and practice, by separating it from psychology and using a fresh, common-sense approach to bring materiality back into the picture.

My core argument for design has been developed by taking one of the main preoccupations of contemporary cultural discourse, the argument for and against the existence of universal truth (posited by many, including Dewey 1934; Fish 1989; and Putnam 1999), and carrying it into the perceptual realm by adopting a pragmatic line of inquiry which questions the very nature of foundational belief. As an alternative, the argument I am constructing is a means of dealing with spatial, visual information that is artistically and conceptually rigorous, making it possible to move debate into the real, tangible world informed by knowledge and ideas.

The problem is philosophical. Endlessly nuanced and variable, the general picture we have of the perceptual process depends on a sensory interface or mode of thinking that somehow intervenes on our behalf to organize various inputs in order to serve intelligence—a "disastrous idea that has haunted Western philosophy since the seventeenth century" (Putnam 1999, 43). As a result, significant chunks of the design process are lost in an arcane, sensory miasma. Although rarely articulated, the concept of a sensory interface is hugely pervasive, affecting almost every facet of Western culture. Henry James described it as a theory that "cheats us of seeing" (quoted by Putnam 1999, 3). What is also evident is that it robs us of artistic sensibility too.

It is difficult to exaggerate how much the general understanding of intelligence is dominated by the false notion of a sensory interface and, in turn, the difficulties it creates. It lies at the heart of the common ideas that art involves a different conceptual framework from science and is a different mode of thinking; that art is a pleasurable pastime whereas science is a serious endeavor; that it is possible to forget all you know in order to appreciate fully a piece of music, a painting, or the landscape, embracing the sensuality of the experience with a clean slate, uncontaminated by knowledge or rationality. It is why, despite so much evidence to the contrary, we still characterize scientists as cool, detached, unencumbered by emotion, and artists as passionate, subjective, and slightly deranged; why we think decisions can be made, on the one hand, intuitively, without knowledge, and on the other, objectively, without value judgments. More generally, it skews the way intelligence is defined or what counts as valid knowledge, and gives a prejudicial and narrow view of the role of language.

Educationally, this is disastrous. For example, at the center of aesthetic experience, the sensory mode of thinking is what students are expected to reap the benefits of if they are to be in any way successful. But the fundamental dichotomy between body and mind enshrined in theories of perception actually creates insoluble puzzles within aesthetics that inevitably spill into design discourse.

Aesthetics, almost more than any other discipline, is dependent on the idea of universal truth. It doesn't seem to matter whether the notion of truth is approached from a transcendental or empirical perspective: in aesthetic theory, what really counts are the universal superstructures that are thought to stand outside culture but also act to underpin and unite our responses.[3] In the attempt to identify these universals, in here, out there, somewhere, we are supposed to set aside all reason, opening ourselves without reservation to what is outside of us in order to sense something "other." However, from a pragmatic perspective, asking anyone to step outside of what he or she knows and to sense significance or beauty as it really is, without the encumbrances of knowledge and culture, is as pointless as it is ridiculous.

Then, most damaging of all, running through a whole range of design theory is the highly pejorative attitude towards the visual, underscoring the supercilious contention that, whatever it is that determines our responses, it is certainly not merely visual. The visual may well be acknowledged as a component, but is also thought to be a distraction. The physical, material qualities of place are thus edged out of the frame because an appreciation of such things is considered too subjective or ephemeral.

As a result, society, generally speaking, has lost the art of critical looking. Through long-term neglect and discrimination, we no longer have the confidence, the appetite, or even the language to talk about appearances. It is abundantly clear, however, that we live and work in a visual, spatial medium. It is both pretentious and foolhardy to think we can manipulate that medium without knowing the implications of what we are dealing with. Undervaluing the cultural and social implications of appearances disables our attempts to understand the impact of the quality of the real, tangible environment on the quality of life.

The Alternative

The alternative is not about rethinking how to design. There are many brilliant designers around who know how to do this only too well. It is more about reevaluating *how we think* we think we design. Using an entirely different basis from which to understand the way we think unlocks a major part of the debate. The philosophical argument is surprisingly simple. Richard Rorty (1982) and other pragmatists insist that it is a mistake to assume there is a difference to be made between modes of thinking in the first place and wrong to think consciousness is fragmented into different kinds of knowing

or intelligence. The characteristics of the subjective/objective dichotomy that theorists and thinkers have been trying so hard to dispel over the last few decades match precisely those of the visual/verbal duality. Both have the same foundational basis. If one dichotomy can be proved erroneous, then so can another. One simple move: take away the foundation stone—the very idea that there are different modes of thinking—and there is no longer any need to worry about how the different realms can be reconciled. All those intractable problems melt away.

Disassemble the argument for different types of reasoning, and the idea that there are different ways of thinking is similarly undone. From a pragmatic perspective, it follows that all thinking, whether in the arts or the sciences, is therefore interpretative and metaphorical; neither uses a special kind of reasoning. Essentially, this is to say that we think the same way no matter what we happen to be thinking about. In understanding emotions or equations, formulae or artistic responses, we interpret, reinterpret, judge, and try to make sense of our feelings because there is simply no other way to make sense of what we see, to make sense of the world. Just because we are looking at a painting does not mean we are thinking in pictures, or that when we are reading a book, we are thinking linguistically. Whatever grabs our attention or catches our eye, no matter what gets us thinking, we always get to think about it by the same route, through language. There are no exceptions, no special cases, ifs, ands, or buts. Language binds us, separates us, it quite literally defines us.

Offering "a middle way between reactionary metaphysics and irresponsible relativism," as Putnam asserts (1999, 5), and redefining the relationship between the senses and intelligence mean that essentially there is no need to choose one or the other. This releases us from the endless debate between positions that are natural or cultural, classical versus romantic, scientific or artistic, theoretical or practical, value laden or quantitative, or approaches that are personal or community-based. Collapsing the visual, intelligence, language, and many other elements of consciousness into a holistic concept of perception takes the supernatural element of design theory and education out of the equation. It also reveals that, far from masking design ability or creativity, concepts and language actually allow us access to the arts, in both their making and criticism.

This radical shift makes it possible to get what Norman Bryson calls "a firm grasp of the tangible world" (1999, 31). As a question of developing a common-sense realism about conception and perception, rather than trusting the world to pass messages to us through sense data, perfect forms, or amenable spirits, we can rely on our reactions and responses being entirely dependent on the sense we make of what we see. We respond to the world through intelligence, and that response is informed by education. Rather than arguing as Arnheim (1986) and many others have done that we should recognize the intelligence of perception, it is better to argue that perception *is* intelligence. This gives us an entirely different mindset to approach design theory and pedagogy.

TRUTH

The challenge of negotiating the territory between the subjective and objective is to work without holding real, unchanging truth to be the ultimate end point of inquiry while, at the same time, to avoid being sucked into the argument that the only alternative to objectivity is to believe everything is relative and dependent on a point of view. Working between the subjective and the objective means to work without the prop of believing there is such a thing as objective neutral truth as much as to avoid dismissing things we don't like/understand/agree with as value judgments, subjective opinions, or questions of taste. However, ditching the idea of universal truth as a guide is not the same as saying that truth does not exist. It is absolutely true that my table does exist; I definitely did have bacon and eggs for breakfast. But

Fig. 3.5. HS2LV conceptual drawing: a visually striking view from HS2 as it enters the Blythe Valley. Designed and drawn by Kathryn Moore, 2012. Used by permission.

these absolute truths are of no particular significance. No matter how scientifically I define a site, all this will ever be is one kind of description—no closer to the essential truth about the site than any other, and as with any other kind of truth, it will also be open to question and subject to change.

EDUCATION

The educational benefits of the philosophical shift I am arguing for are considerable. If we just look at a couple of topics, the extent of this reconceptualization will become evident. Rather than being a particular mode of thinking, visual skill is an educated awareness of the traditions of the landscape as well as its physical materiality. It is about having a strong sense of our culture, neither generic

nor archetypal, but a learned, cultivated skill, comprising observation and discernment within the traditions, materiality, and ideas of a particular medium. It is a truly critical component of artistic sensibility. To consider the visual in this light makes it possible to learn about why landscapes look the way they do, how and why we respond to places, and then to apply this knowledge to design. When you first see the Manhattan skyline or the Statue of Liberty, for example, the impact is so intense because of the associations gleaned from numerous books, films, and anecdotes. These influences flood in because we recognize *directly* the physical fabric of what we see, the spatial, visual qualities of landscape, its form and character, its myths and legends (fig. 3.5).

What might cause an aesthetic experience,

whether an object or activity, is immaterial. It is the response that counts. If we are lucky, we will be rendered speechless by the power of a painting, the beauty of a landscape, or the balance of a mathematical equation. Pleasurable, emotional, moving, and inspiring experiences can happen when reading a classic novel or merely wandering down the street—it is not the painting that counts, or the landscape or the music, but the quality of the experience. Since the quality of the experience we have is defined by what we know, this makes it entirely accessible. Knowledge we can teach; judgments and values can be learned.

Having startling consequences for the roles of language, the emotions, intelligence, and subjectivity, understanding the value of aesthetic experience also changes our concepts of objectivity. The bottom line is that, as individuals or as a community, in any study, design or otherwise, we are constrained or liberated by the language and concepts we have at our disposal. There is no other way of knowing, no other kind of meaning to uncover, no "genius loci" to give us a nudge in the right direction. Neither the site nor what lies beneath, within, or without it, nor even the fears and desires of our prehistoric ancestors can speak to us beyond what we know.

There is no way to operate with the presumed objective neutrality of a so-called scientific approach. We need a healthy measure of skepticism to deal with the "hard facts" enshrined in regional spatial plans, perennially used to justify the economic imperative for new roads, the distribution of new settlements, how big they should be or the cost the market will stand in terms of quality housing or town center development. The evidence of the impact of such quantitative factual decisions is only too evident. Just look around any town or city. Or compare today's transport, housing, and agricultural policies with those of fifteen years ago. Were all those experts just plain wrong back then, or were they simply working under different circumstances and with a different set of values—different ideals?

Reevaluating fundamental assumptions about objectivity inevitably has an impact on our pedagogy, especially in teaching aspects of the design process. For example, rather than simply aiding the mechanical or practical part of a project, we have learned to see technology as a means of understanding how far materials might be pushed or manipulated in order to express ideas with style and confidence. Similarly, drawing, rather than just a technique offering access to intuition, somehow kicking part of the brain into touch, should be valued more as a way of working things out, exploring ideas, and speculating about possibilities. As an investigative tool, drawing is hard to beat.

Reevaluating our pedagogy on technology and drawing becomes a matter of realizing that "'meaning' is not simply something 'expressed' or 'reflected' in language: it is actually *produced* by it" (Eagleton 1983, 60). There is no other way for meaning to be manifest. This is the case in all experience. Reevaluating such fundamental assumptions sets a different kind of agenda for theory, too. Notwithstanding if theory is esoteric or down-to-earth, jargon-laden or straight-forward, there remains an urgent need to refocus and develop theory that is insightful enough to inform the processes and the values of design. This new kind of theory would not be particularly concerned with whether an approach is semiotic, structuralist, or post-structuralist, phenomenological, or even hermeneutic. There would be no need to fight one philosophical corner against another, and it would be unnecessary to depend on the language or jargon of that discourse in order to signal allegiance or make a point. The main purpose of our teaching transcends all that: to delve into the particularities, appropriateness, and expression of certain ideas in built form, given the place, time, and context.

FORM

Recognizing that both perception and language are interpretive removes a blindfold.

This recognition is the final radical shift that enables us to understand one of the most obscure aspects of the whole design process, that of *generating form*, demonstrating the indivisibility of ideas, theory, expression, and technology in practice, realizing that it is as impossible to design without concepts as it is to talk without a tongue. Sensible discussions can emerge about the making of informed, imaginative, and often-difficult design decisions, making it clear that there is nothing magical or mysterious about all this.

The impact this makes in the studio is in many respects quite simple—it is a matter of consistently asking "why" things look like they do, what ideas are being worked with, what they look like, and why they are appropriate given the site, project brief, and context. What spatial principles are being worked with, and how are ideas being expressed—whether at a strategic or a detailed level. It means asking for an explanation as to why a particular kind of materiality is involved (light, shape, form, texture) and what quality of experience is being shaped for what kind of user. This way of teaching is about encouraging students to plunge wholeheartedly into the visual, spatial world, working with ideas, space, form and materiality. It is also about inserting moves and tactics to discourage the habitual or clichéd decisions that students often fall back on (presumed to be instinctive or intuitive

Fig. 3.6. Detail of city diagram (Kidderminster) produced for postgraduate program, Birmingham Institute of Art & Design. Aiming to reestablish the heart of Kidderminster, and forming a reference point throughout the design process, it acts as a reminder of the need to establish Kidderminster as a setting to host life events, ceremonies, and seasonal festivals, and as a major tourist destination. Designed and drawn by Hannah Leonard, 2011. Used by permission.

but actually learned) instead of pushing and challenging students to go further. This develops their confidence in knowing which line of inquiry is a good one to follow, supports risk-taking, and challenges both faculty- and student-held preconceptions. Above all else, to teach in this way is *not* to presume that anything visual or spatial or conceptual is self-evident. It *is* to enter an ambiguous world (fig. 3.6).

Democratic Education

Dispensing with the metaphysical dimension of perception introduces a much-needed critical element to all of design education, not just certain parts of its theory or technology, and helps to democratize the process. From a student's point of view, she can at last stand up and lead the critic and, to a certain extent, take control of her own learning. It shifts the balance of power a little. Instead of being left with the sinking feeling that they must be missing that indefinable "something," students are able to start working confidently with ideas, expressing and adapting them within the medium or brief. They can be assured that advancing their design skill is a question of learning—not voodoo. No one is suggesting that every student will become a brilliant designer, but at least everyone stands an even chance of learning *how* to design. This has to be better than being led to believe that design is an inherent ability, a talent, or "gift" you might never manage to open.

SHAPING THE QUALITY OF EXPERIENCE

This approach has wider social and political implications and demands. Moving the debate away from the arcane and unknowable into the real world informed by knowledge and ideas serves to remind us how much theory and philosophy can learn from practice, rather than the reverse. Seen from this perspective, landscape is not just about ecology, nature conservation, or matters of heritage. It is not only about the physical context, the constructed public realm, the national parks, coastlines, squares, promenades, and streets. Landscape also reflects our memories and values, the sense of pride we share in the places where we work and live, the experiences we have of a place, as citizens, employers, visitors, students, and tourists. It is the material, cultural, and social context of our lives.

The landscape is about ideas, and the expression of these ideas shapes the quality of our experience. Rather than ideas versus nature, we have ideas of nature. Instead of seeing nature as something separate from culture, from ourselves, we must recognize that in the way we live our lives, with every intervention we make, we are expressing (consciously or not) an attitude toward the physical world. The choice is not whether we work with art or ecology, with nature or culture, but how considerately, imaginatively, and responsibly we go about our business, because for every one of our actions, there is a reaction in the physical world. We make an impact on the world every minute of every day of our lives. Working with natural processes, given the global challenges we face, is an ecological imperative. We have no choice in the matter. But it is the whole thing, the ideas and values we hold and their expression in physical form, be it green, gray, or blue, that defines us. This is what frames the experiences we all have of the places we live in, and these experiences form the basis of a properly relevant definition of nature. After all, natural systems don't stop where the buildings start.

Shifting paradigms from the metaphysical to the pragmatic enables us to understand more about the nature of critical visual sensibility and the role it plays in design. It does not mean that mystery and ambiguity no longer have a place, or even that designs cannot be inspired by metaphysical concepts. But it does mean that we can remove some of the mystery from the actual process of designing. Visual skill can be seen as something that we need to learn in order to become designers. Dewey's suggestion that aesthetics is "a manifestation, a record and celebration of the life of a civilization, a means of promoting its development, and is also the ultimate judgment upon the quality of a civilization" (1934, 326), hands us the responsibility to ensure the record is a good one.

RESEARCH

Seen from this perspective, design is a question of investigation, research, and decision-making. Designing is about making propositions, presenting a vision for the future. Central to the discipline is the forward-thinking, the anticipatory, and predictive nature of its practice. And as far as research is concerned, with any and every part of the design process transparent and accessible to investigation, it is also clear that the limits of our inquiries are governed only by our knowledge, values, and experience. Responsibility for understanding what sense we make of the world is handed back to us. The driest, most reductive statistical equation or number-crunching analysis is just as full of values, presumptions, and preconceptions as any ephemeral, instinctive, subjective response. Simply look at the ongoing debates relating to climate change to see how easily facts may be manipulated by interpretation, often limited or motivated by values held by the interpreter. This doesn't even begin to address the chore of finding any consensus as to what is an actual fact and what isn't.

We should recognize that what is considered to be clear and rigorous research is absolutely contingent upon the knowledge, values, and opinions of those who judge it. This explains why Swaffield and Deming (2011) find that what is valued in research is shaped by academic location, the educational background of academics, and the particular approach of editors and reviewers. Those undertaking research effectively enter a lion's den, and work can easily end up in the hands of someone with a conflicting agenda or an entirely different view of the world. So as supervisors, reviewers, and editors, our role is to be informed and make judgments from a position of knowledge and experience, aware of our prejudices, preconceptions, and desires. The hard part is first to recognize what these are and then to have the courage to put them to one side if necessary. This is being properly objective—not trying to gauge how closely the work measures up to our own ideas but being open and pragmatic enough to appreciate what might be an entirely different way of understanding things (Rorty 1999, 181).

Conclusion

In order to overcome a long period of technological stagnation in landscape architecture, and rather than staying within the safety of fixed disciplinary parameters, we need to be more aggressively expansive. This requires us to appropriate and operate more confidently in making connections among disciplines, linking theory and practice, ideas, and form, evaluating the ethical, aesthetic, ecological, and artistic value of the physical and imagined environments with the explicit purpose of investigating how this knowledge can be used directly to inform design.

Changing the focus in the way we think of landscape from technology towards ideas, seeing the landscape as both a cultural *and* natural resource and a physical *and* abstract entity, having economic *and* social value, looking at the experience people have of their physical environment as well as making the vital connections between governance, culture, health, and economics—these steps go some of the way to provide viable new platforms from which to deal holistically with the rural and the urban, wilderness and man-made, the most treasured and memorable and as well as the unloved and degraded. Setting a new agenda for research will bring fresh insights to shape the future of our environment.

Belief in the notion of the sensory interface is so endemic that it has practically become a fact of life. There is, therefore, a degree of anxiety to contend with, questioning old certainties in order to realize the interpretive, transient nature of everything we believe to be true. Shifting any inquiry away from the unequivocal towards the ambiguous is perhaps one of the most difficult aspects of this emerging pragmatic paradigm. It is not just another way of saying that anything goes; rather, it suggests that our work must now be judged against different criteria.

Truth is contingent, sometimes enduring, but never immutable. Beliefs change. And it is this flexibility that gives us such a great opportunity. If we have the confidence to move away from the central hard core of scientific assumption and methodology, there is a real chance to develop new approaches, make connections across and among disciplines, erase rigidly drawn boundaries delineating and distinguishing practice from theory. The old Cartesian duality is a house of cards—time to blow it down.

NOTES

1. This chapter was developed based on extracts from previous lectures and published material (Moore 2010).

2. In addition to being featured in *Topos* magazine (March 2013), project drawings by Kathryn Moore have been featured in a national touring exhibition opening in London, March 2013, rhythm-presence .co.uk/#home. The HS2LV proposal has also been presented at UNESCO, United Nations, and other professional conferences to illustrate some of the potential practical implications of an international landscape convention.

3. Dependence on this belief is exposed in the dialogue between Richard Rorty and Richard Shusterman (Rorty 2001); also see Terry Eagleton (2003).

REFERENCES

Arnheim, Rudolf. 1986. *New Essays on the Psychology of Art*. Berkeley: University of California Press.

Bryson, Norman. 1999/2001. "The Natural Attitude." In *Visual Culture: The Reader*, ed. Jessica Evans and Stuart Hall. London: Sage Publications in association with The Open University.

Dewey, John. 1934. *Art as Experience*. New York: Berkley (reprinted 1980).

Eagleton, Terry. 1983. *Literary Theory: An Introduction*. Minneapolis: University of Minnesota Press.

———. 2003. *After Theory*. London: Allen Lane.

Fish, Stanley. 1989. *Doing What Comes Naturally: Change, Rhetoric and the Practice of Theory in Literary and Legal Studies*. Durham, NC: Duke University Press.

Moore, Kathryn. 2010. *Overlooking the Visual: Demystifying the Art of Design*. Abingdon, UK: Routledge.

Putnam, Hilary. 1999. *The Threefold Cord: Mind, Body, and World*. New York: Columbia University Press.

Rorty, Richard. 1982. *Consequences of Pragmatism: Essays, 1972–1980*. Minneapolis: University of Minnesota Press.

———. 1999. *Philosophy and Social Hope*. London: Penguin.

———. 2001. "Response to Richard Shusterman." In *Richard Rorty: Critical Dialogues*, ed. Matthew Festenstein and Simon Thompson, 153–57. Cambridge, UK: Polity Press.

Swaffield, Simon, and M. Elen Deming. 2011. "Research Strategies in Landscape Architecture: Mapping the Terrain." *JOLA* 6.1 (Spring), 34–45.

CATHERINE SEAVITT NORDENSON

De-domestication and the Wild

De-domestication is the process of establishing a formerly domesticated species, either plant or animal, in the wild (Gamborg et al. 2010). Often this process involves some change at the genetic level and is similar to the artificial selection and breeding of domesticated animals and plants by humans. De-domestication is considered successful when the animals are self-sustainable in the wild without supplemental feeding or human care. Recently, de-domesticated plants and animals have been studied as tools for landscape restoration—for example, consider the genetic development of new varieties of perennial wheat species proposed by Stan Cox and other research scientists at the Land Institute, a private research organization in Salina, Kansas (Cox 2008), as a strategy for revitalizing and sustaining midwestern agricultural grasslands. Similarly, the de-domestication of herbivores and the impact produced by wild herds has been identified as a useful strategy for certain practices of landscape restoration (Klaver 2002). Animal impact by de-domesticated cattle and horses includes short-duration but high-impact trampling, grazing, and dunging, achieved as the herds wander within an open territory without the limits of managed pasturing. This technique is particularly effective for the revitalization of grasslands and the prevention of forest succession; the grazers turn the earth and inhibit the growth of shrubs and trees.

Fig. 4.1. The Oostvaardersplassen, Netherlands. View of Heck cattle and konik horses, 2005. Photo by Gerard Meijssen. Wikimedia Commons, CC BY-SA 3.0.

Surprisingly, the use of de-domesticated animals as tools in anthropogenic restoration practices has not led to many discussions of their ethical treatment. Although considered wild, these de-domesticated animals serve a particular function as a tool of human-planned processes. What is the responsibility of humans to these "working" animals? At what point should humans intervene in the care of a de-domesticated herd—for instance by providing veterinary assistance, supplemental feeding during times of scarcity, population control, and the prevention of suffering during the death of an individual animal? What are the ethical implications given natural predation by carnivores (for example, by successfully reintroduced wolves) upon herds of de-domesticated herbivores who may no longer possess the same defense mechanisms or skills of their wild ancestors? And, perhaps more fundamentally, what are the ethics of the genetic manipulation of animals returned to a wild state, the process of so-called "back-breeding," which originated in the early twentieth century in both Germany and Poland?

These ethical questions, and the intentional creation of a new, "natural" territory shaped by the grazing habits of these animals, suggest a rereading of Aldo Leopold's 1949 essay, "The Land Ethic." Leopold states: "All ethics so far evolved rest upon a single premise: that the individual is a member of a community of interdependent parts. . . . The land ethic simply enlarges the boundaries of the community to include soils, waters, plants, and animals, or collectively: the land" (239). The relationships among the members of this community, as seen through the lens of this newly created land, have become increasingly complex. "Nature," or a new "wild" land, is generated through a process initiated by humans yet shaped by animals.

The understanding of the land ethic must be radically repositioned.

Oostvaardersplassen Nature Reserve

To address these questions, this chapter examines a case study of de-domestication in the Oostvaardersplassen, a nature reserve in the Netherlands. This reserve, twenty-three square miles in area, currently has the largest population of wild Heck cattle (*Bos taurus*) in the world, as well as a large population of the wild konik horse (*Equus ferus caballus*) (fig. 4.1). The Oostvaardersplassen is part of a large engineered polder—a fully artificial landscape reclaimed from the sea. Franz Vera, the Dutch ecologist who first brought the Heck cattle and konik horses to the Oostvaardersplassen in the early 1980s, argues that the landscape of the Oostvaardersplassen mimics an ecosystem quite similar to early European wetland marsh areas that would have supported a population of wild herbivores (Kolbert 2012; Vera 2000).

What are the origins of this unlikely "primeval" landscape? The Oostvaardersplassen is located along the edge of a lake in the Netherlands known as the IJsselmeer. Once called the Zuiderzee, an inlet of the North Sea, this water body was transformed into a freshwater lake in 1932 through the construction of a dike across the mouth of

the inlet (fig. 4.2). The connection to the sea was blocked.

Later, the IJsselmeer was further transformed through the construction of more dikes and draining. Reeds were planted on the old sea bottom, drying out the soil. The Netherlands gained approximately 626 square miles of reclaimed land through the creation of three large polders along its shores between 1942 and 1968. In 1973, a fourth polder was diked but not drained (Campbell 1999). By the late 1980s, the bottom of the former inlet had thus been transformed into land. Much of the reclaimed land is used for agriculture, but industrial uses and planned new towns also developed, particularly the

town of Lelystad, founded in 1966 in the East Flevoland polder, just northwest of the Oostvaardersplassen. Completed in 1968, the future Oostvaardersplassen polder within South Flevoland was intended to be developed as an industrial site, but the area was abandoned and "nature" began to reclaim the site.

The area became an attractive freshwater wetland destination for migratory birds, particularly the Greylag goose (*Anser anser*). Their molting and grazing impact created a patchwork of areas of open water and marsh plant growth (fig. 4.3). The current perimeter of the site was delineated and established as a nature reserve in 1982, and, in 1996,

Fig. 4.3. Konik horses and shorebirds at the Oostvaardersplassen, 2008. Photo by E. M. Kintzel, I. Van Stokkum. Wikimedia Commons, CC BY-SA 3.0.

OPPOSITE:
Fig. 4.2. IJsselmeer, Netherlands, 2010. The location of the Oostvaardersplassen polder is outlined in white. Note location of the 1932 linear dike across the top middle of the image. Amsterdam is southwest of the Oostvaardersplassen. U.S. Geological Survey.

Fig. 4.4. Sluice for water-level management at the Oostvaardersplassen, 2005. Photo by Gerard Meijssen. Wikimedia Commons, CC BY-SA 3.0.

management of the site was transferred to the Staatsbosbeheer, the Netherlands' State Forestry Management Service. Representatives of the Staatsbosbeheer argue that the size of the twenty-three-square-mile Oostvaardersplassen nature reserve—approximately equal in area to the island of Manhattan—is large enough to function as its own natural ecosystem. However, like all polders, its hydrology is carefully controlled through a manual sluice system (fig. 4.4). In addition, the site is fully fenced and closed to public access except for a small area. Thus,

despite the claim that the site operates as a "natural" ecosystem, it is, in fact, a highly artificial and controlled site.

In the early 1980s, the Staatsbosbeheer, with the counsel of Franz Vera, chose to introduce several large de-domesticated herbivores to the Oostvaardersplassen as a management and planning strategy. In 1983, thirty-two Heck cattle from Germany were introduced. In 1985, twenty konik horses from Poland were introduced. And in 1992 and 1993, fifty-seven wild red deer (*Cervus elaphus*) from Scotland were introduced

(Oostvaardersplassen 2012). The intention was for these herbivores to trample and graze the polder's emergent grasslands as wild herds so that the successional willow growth would not lead to forestation that would destroy the existing wetland marsh. The wild herds are provided with no winter shelter, receive no supplemental feeding, and are given no veterinary assistance (ICMO 2006). There are no large top predators (such as the wolf or lynx) in the Oostvaardersplassen, so population control of these large herbivores still occurs at the level of the human predator—the ranger. Since 2006, the reserve has instituted a "reactive policy" (in contrast to a "pro-active policy"). According to this policy, animals that have separated themselves from the herd due to injury or disease may be shot, in theory to relieve them of suffering "while they are still capable of standing" (ICMO 2006, 10). The stated goal of the reserve is to shoot 90 percent of all dying large herbivores in this condition (ICMO 2006). In a site as large as Manhattan, one can imagine that this management directive must be difficult to achieve.

The reserve conducts an aerial population survey every three years; in 2008 the total population of large herbivores within the refuge, including Heck cattle, konik horses, and red deer, was estimated at approximately four thousand animals (Oostvaardersplassen

2012). Winters can be harsh at the Oostvaardersplassen and have proven very difficult for the survival of large herbivores. Recently there have been large winter die-offs, raising ethical questions (ICMO 2006), particularly during the cold winters of 2004–5 and 2009–10 when over 25 percent of the large herbivores died (ICMO2 2010). Yet even without large carnivorous predators at the site, the Staatsbosbeheer argues that many types of organisms benefit from these herbivores' cadavers—including bacteria, carrion beetles, foxes, ravens, and white-tailed eagles.

Conditions at the Oostvaardersplassen nature reserve raise three ethical issues for landscape architects, ecologists, and humans in general: the environmental ethics of human creation of a "natural" wilderness through the shaping of ecological systems by animals; the animal ethics of management of these herbivores, as individuals and as a herd, within such nature reserves; and the genetic ethics revealed in the broader historical question of de-domestication, sometimes called back-breeding, of these large wild herbivores. As the anthropocene footprint expands worldwide, eliminating any notion of an untouched "wilderness," the importance of considering the role of humans in the ecological restoration or intentional design of new wildlands as well as the ethics of our relationship with the nonhuman biota within

these territories must be carefully considered. The issues raised by the Oostvaardersplassen reserve are important to understand, as they resonate with many contemporary projects currently underway in Europe and North America: wildland territories and ecological corridors repopulated with migrant species, wildlife refuges supporting the habitat of endangered species, and increasingly popular private hunting grounds stocked with exotic game.

Fig. 4.5. Painting of a prehistoric aurochs in the caves of Lascaux, France, upper Paleolithic period. Photo by Fernand Windels, 1949.

The Aurochs and Heck Cattle

The prehistoric aurochs (*Bos primigenius*) is a species of wild bovine originating in Europe and Asia from which all domestic cattle are thought to descend. Its image appears in the prehistoric cave paintings of Lascaux; the aurochs was considered a potent symbol of power and freedom in many cultures (fig. 4.5). Domestication of aurochses (plural form) occurred parallel to the development of farming, approximately ten thousand years ago. The aurochs was huge, often almost six feet tall. Through livestock husbandry and selective breeding, wild cattle gradually became smaller, more docile, and were articulated into many varieties or breeds. Widespread hunting of the aurochs across Europe and Asia, as well as its banishment from territories claimed by domesticated

Fig. 4.6. Three
Heck cattle at the
Oostvaardersplassen.
Photo copyright © 2011
Virginie Monchy, all
rights reserved. Used by
permission.

cattle and human development, led to its eventual extinction in 1627 on the royal hunting grounds of the Polish king. Today, the former presence of the aurochs is maintained merely as an artifact within museums.[1]

In the early 1920s, two zoologist brothers, Heinz Heck, the director of the Munich zoo, and Lutz Heck, the director of Berlin's zoo, began two separate breeding experiments in an attempt to recreate the wild aurochs. Although modern genetics has confirmed the impossibility of the recreation of an extinct species through the breeding of existing species with similar physical characteristics, Heinz Heck claimed in 1932 to have resurrected the aurochs with just a twelve-year "back-breeding" process. "Der Ur lebt wieder" ("the Aurochs lives again") was Heinz's triumphant cry upon the birth of a bull named Glachl (van Vuure 2005, 339). This animal was bred by the crossing of various European cattle breeds, including German, Scottish, and Ukrainian cows and bulls, whose physical characteristics and appearance Heck considered similar to those of the extinct aurochs.

This notion of pure race resonated with the contemporaneous ideology of Nazi Germany, whose leadership was interested not only in a pure race of humans but also in studying the possibility of resurrecting genetically pure European animals. Hitler's designated successor, Hermann Göring, the head of the German Air Force and an avid hunter, was particularly supportive of the work of the Heck brothers, as he imagined them stocking his large hunting reserve at the Rominten Heath, in what was then northeast Prussia, with wild primeval beasts, thereby creating a unique royal Nazi hunting menagerie (Milmo 2009). The selection and breeding project ended with World War II in 1944, and only a small number of the stock of Heck cattle from the Munich Zoo survived the war (van Vuure 2005). It is assumed that some postwar interbreeding may have occurred, but small populations of the Heck did continue in West Germany. Currently, the largest concentration of Heck cattle descendants, over four hundred head, live in the Netherlands' Oostvaardersplassen (fig. 4.6), introduced to the site from Germany by the ecologist Franz Vera (Kolbert 2012).

The Tarpan and the Konik Horse

The Tarpan (*Equus ferus ferus*) is the equine equivalent of the aurochs, a wild horse once roaming Europe and Asia. The image of a Tarpan-like horse appears in the Lascaux cave drawings; unlike the aurochs, which was extinct by 1627, the Tarpan survived in the wild in central Europe until the late

Fig. 4.7. Painting of a prehistoric Tarpan in the caves of Lascaux. Photo by Fernand Windels, 1949.

nineteenth century (fig. 4.7). Loss of habitat and overhunting caused their number to diminish until only a small number remained in the Bialowieza Forest, an ancient woodland straddling the border of Poland and Belarus that was once a royal Polish hunting ground. The last captive Tarpan horse died in a Russian zoo in 1909.

In the early 1930s, the Heck brothers initiated another back-breeding project to recreate this prehistoric wild horse, drawn by the rich folklore of the German forest hunting culture. The product of this experimental

program, which bred Icelandic and Gotland mares, domesticated horses of small stature, with the Przewalski stallion (*Equus ferus przewalskii*), a wild horse native to the steppes of central Asia, was the Heck pony, a colt born in 1933.

A second Tarpan back-breeding and restoration program was begun in 1936 by a Polish professor, Tadeusz Vetulani, of the Poznan University. Vetulani argued that there was a forest variety of the Tarpan that had developed independently of the type living in the steppes of Eastern Europe. This

"local" forest Tarpan had survived into the mid-eighteenth century in forested areas of Poland, Lithuania, and Prussia. Vetulani attempted to recreate and revive this Tarpan type through a selective breeding program that drew from the descendants of the local peasants' horses. The Polish government gave a show of support by commandeering for Vetulani all of the horses that displayed Tarpan-like features, including a dun-colored coat, dark face mask, and a dorsal stripe, and establishing a nature preserve specifically for them in the forests of Bialowieza, Poland. This selectively bred horse was called the konik horse (*Equus ferus f. caballus*). *Konik* is the Polish diminutive for horse; this breed is often called the Polish pony or Polish primitive horse. These are strong and hardy horses, capable of foraging successfully in the wild. The konik is found in Germany, Belgium, Latvia, and England. And in 1985, the horse was introduced by Vera to the Oostvaardersplassen in the Netherlands, which now has one of the world's strongest populations of the konik horse (fig. 4.8).

The Ethics of the Nature Reserve: New Nature

The genetic specifications of the characteristics of both the konik horse and Heck cattle have led to their implementation as a maintenance tool within a specific landscape.

In the case of the Oostvaardersplassen, these herbivores' grazing habits in the wild—including high-impact trampling and grazing—are applied to reduce the progression of forestation in this grassland nature reserve. The animals, in a sense, have become part of a human-engineered design palette. Much as one would specify a plant list for a site, these de-domesticated animals, with their specific grazing habits, are selected for their impact on a particular terrain.

What are the ethics of designating a territory as "wild" and inserting within it a maintenance program driven by the particular characteristic behaviors of animals? Is releasing this territory to the animals a way of establishing a "new nature" that is determined by the animals themselves? Or does it produce a variant of the classic climax community model, a steady state of an idealized ecosystem within a contained territory? What then is the role of the human in this particular land ethic, or is the ultimate goal of such a new nature the complete elimination of human management? If the goal of this new nature is to create a self-sustaining wild system, the enclosed patch inhabited by a migratory herbivore species seems non-sustainable. But what are the ethics of introducing a meat-eating predator, such as the wolf or lynx, to a fully enclosed, vegetarian ecology, currently a paradise

for de-domesticated herbivores, with no predation except euthanasia?

Richard T. T. Forman (1995) has long advocated a continuous corridor-and-patch ecology, a system that might sustain ecological systems within the fabric of human development. Considering proposed future plans for the Oostvaardersplassen, it appears that the introduction of large carnivores certainly could occur, whether intentional or not, through the opening of a proposed migration corridor connecting the nature reserve to an established national forest to the south. This link from the Oostvaardersplassen to the Horsterwold national forest, on the southern edge of the same South Flevoland polder, via an ecological corridor, is currently being planned by the Staatsbosbeheer (ICMO 2006). This proposal is called the Oostvaardersland and would result in a combined nature reserve area of over fifty square miles, more than doubling the current size of the Oostvaardersplassen (fig. 4.9).

On an even grander scale, environmental commissioners in the European Union launched an initiative entitled Natura 2000, under the 1992 Habitats Directive, as a measure to improve biodiversity in Europe. Natura 2000 would link individual nature reserves together to create a network of conserved lands; these would protect

Fig. 4.9. A proposal for Oostvaardersland, an expansion of the Oostvaardersplassen nature reserve to the Horsterwold forest via a migration corridor, 2006. Photo from International Committee on the Management of Large Herbivores in the Oostvaardersplassen, Staatsbosbeheer/Wing Process Consultancy, June 2006.

threatened species and their habitats as well as improve biodiversity. This would open up the possibility of even longer migration corridors for the large herbivores of the Oostvaardersplassen (Oostvaardersplassen, 2012). With the implementation of Natura 2000, the Oostvaardersland has the potential to gain a connection to the Veluwe forest and massif, and beyond to Germany.

An unfencing of the Oostvaardersplassen, along with the relative success of recent European initiatives that support the reintroduction of large carnivores such as the wolf (*Canis lupus*) and Eurasian lynx (*Lynx lynx*), would create quite a different "wildland" terrain. It is unclear how the de-domesticated herbivores would fare in this system—would a back-bred animal have the defensive capability needed to participate in this newly dynamic system, without the intercession of humans? What are the human obligations for the ethical treatment of animals—should these de-domesticated animals be considered individuals, thus focusing on the care of each animal? Or are they now to be considered under a rubric of the ethical management of wildlife, with an emphasis on the well-being of the species group? Perhaps the position of the Dutch biologist and ethicist Jac A. A. Swart, called "non-specific care" (2005, 258), is the most effective one, in which human responsibility for wild animals is seen in relationship to both the health of the species group and the health of their natural environment. Yet the discussion of the health of the "natural" environment is still fraught, particularly when the nature in question is a managed reserve. Which environment might be considered healthier (or a more ethical nature reserve) for a de-domesticated herbivore: the "new nature" of the Oostvaardersplassen, with its lack of large predators, or a future Oostvaardersland or Natura 2000, in which predators would certainly enter?

Conclusions and Reflections: Faking the Wild?

The fascinating case of the Oostvaardersplassen raises many ethical questions, not only for the future management of this particular nature reserve but also for the landscape architects, restoration ecologists, and conservation biologists who design and manage similar reserves. Perhaps one of the most pertinent and difficult questions, particularly as more migratory species are reestablished within wildland corridors in both Europe and North America, is the notion of this human-initiated recreation of "nature" and the wild. The ethical imperatives of humans toward animals, in this case the de-domesticated herbivore, are inseparable from those of humans toward the terrain inhabited by these animals. In other words, the ethics of animal treatment are intertwined with the ethics of the land. The environmental philosopher and ethicist Richard Elliot (1982) and others have argued that human restoration of the land is merely "faking nature."

Beyond even ecological restoration, what are the possibilities and the implications of faking the wild? Is the human effort and management required to establish a facsimile of an imagined primeval wilderness warranted? What is our responsibility toward animals, particularly in this case, toward the large herbivores being used as management tools? In enforcing a policy of euthanasia, are we accurately simulating the processes of "wild" nature and, if so, does that make these actions ethically reprehensible or laudable? Does the romanticism of the idea of recreating a place that reflects a time before human impact paradoxically require even more human domination? Finally, as humans, are we stewards of the environment, or just another animal species operating within an environment?

Revisiting Leopold's "Land Ethic," we see that his notion of land is not at all a terrain created by human action—he posits a wilderness that predates human impact. Yet our anthropocene footprint on the land is vast, such that even this conception of nature and the wild needs redefinition. The philosopher and anthropologist Bruno Latour (1991) may provide a more workable thesis via his notion of a human/nonhuman collective, or a newly complex community of parts or "parliament" in which we as humans are obliged to give voice and representation to the nonhumans within this democracy. Humans are the potential designers of this construct of a new nature, and of new wild lands; we must establish a way of designing with both plant and animal systems, as well as their extensive terrains, in a more ethical, complex parliament. The ecological theory of the self-perpetuating climax community model of plants and animals may no longer be a viable model in this new nature (Bratton 2000). A dynamic, flexible, and resilient ecological system with an ethical place for all within the community, and the support of biodiversity at its core, must be embraced by humans.

NOTES

1. The right horn of the last aurochs bull, which died in 1620, was transformed into a commemorative hunting horn and is now in the Livrusthammeren in Stockholm. An intact aurochs skeleton, discovered in a Swedish peat bog, is displayed in the National Museum of Denmark. Aurochs skulls and horns are displayed in several European museums.

REFERENCES

Bratton, Susan Power. 2000. "Alternative Models of Ecosystem Restoration." *Environmental Restoration: Ethics, Theory, and Practice*, ed. William Throop, 53–70. Amherst, NY: Humanity Books.

Campbell, Robert Wellman, ed. 1999. "IJsselmeer, Netherlands: 1964, 1973, 1987." In *Earthshots: Satellite Images of Environmental Change*. Washington DC: U.S. Geological Survey. earthshots.usgs.gov (accessed January 15, 2012).

Cox, Stan. 2008. "Ending 10,000 Years of Conflict between Agriculture and Nature." *Science in Society* 39 (Fall): 72–75.

Elliot, Robert. 1982. "Faking Nature." *Inquiry* 25 (1): 81–93.

Forman, Richard T. T. 1995. *Land Mosaics: The Ecology of Landscapes and Regions*. Cambridge, UK: Cambridge University Press.

Gamborg, Christian, et al. 2010. "De-domestication: Ethics at the Intersection of Landscape Restoration and Animal Welfare." *Environmental Values* 19 (1): 57–78.

ICMO (International Commission on Management of the Oostvaardersplassen). 2006. "Reconciling Nature and Human Interests: Advice of the International Committee on the Management of Large Herbivores in the Oostvaardersplassen (ICMO)." *WING Report* 18 (June), The Hague/Wageningen, Netherlands. wildexperiments.files.wordpress.com/2011/05/icmo-2005.pdf (accessed January 15, 2012).

ICMO2 (Second International Commission on Management of the Oostvaardersplassen). 2010. "Natural Processes, Animal Welfare, Moral Aspects and Management of the Oostvaardersplassen: Report of the Second International Committee on the Management of Large Herbivores in the Oostvaardersplassen (ICMO)." *Wing Report* 39 (November), The Hague/Wageningen, Netherlands. www.wing-wageningen.nl (accessed January 15, 2012).

Klaver, Irene, et al. 2002. "Born to Be Wild: A Pluralistic Ethics Concerning Introduced Large Herbivores in the Netherlands." *Environmental Ethics* 24 (1): 3–26.

Kolbert, Elizabeth. 2012. "Recall of the Wild." *New Yorker* (December 24 and 31), 50–60.

Latour, Bruno. 1991. *We Have Never Been Modern*. Trans. Catherine Porter. Cambridge, MA: Harvard University Press.

Leopold, Aldo. 1949. *A Sand County Almanac and Sketches Here and There*. New York: Oxford University Press.

Milmo, Cahal. 2009. "Hitler Has Got Only One Bull." *Irish Independent News* (April 22).

Oostvaardersplassen. 2012. *Staatsbosbeheer, The Netherlands*. www.staatsbosbeheer.nl/English/Oostvaardersplassen/ (accessed January 15, 2012).

Swart, Jac A. A. 2005. "Care for the Wild: An Integrative View on Wild and Domesticated Animals." *Environmental Values* 14 (2): 251–63.

van Vuure, Cis T. 2005. *Retracing the Aurochs: History, Morphology, and Ecology of an Extinct Wild Ox*. Sofia, Bulgaria: Pensoft Publishers.

Vera, Franz W. M. 2000. *Grazing Ecology and Forest History*. Oxford, UK: CABI Publishing.

Windels, Fernand. 1949. *The Lascaux Cave Paintings*. London: Faber and Faber Ltd.

KYLE D. BROWN

Toward Ecological Sovereignty

THE REGENERATIVE COMMUNITIES INITIATIVE

Environmental degradation has created challenging conditions in many urban communities, particularly those characterized as poor, minority, and otherwise marginalized. These conditions have contributed to significant public health challenges in many such communities in Los Angeles County, including disproportionately high rates of obesity, diabetes, heart disease, and asthma. The Los Angeles County Department of Public Health recognizes this relationship between poor public health and marginalized communities. They compare health data with an "Economic Hardship Index" (County of Los Angeles 2010, 3) that considers factors such as household size, income, prevalence of poverty, unemployment, lack of educational attainment, and preponderance of youth and elderly, and have found a significant correlation between economic hardship and low life expectancy. Communities with a high economic hardship index rating are more likely to face health challenges connected to environmental conditions.

Many academic institutions have engaged marginalized communities in an effort to address these conditions, often assuming the role of community design consultant, action-researcher, or community facilitator. Since 1994, the Lyle Center for Regenerative Studies, an interdisciplinary unit at California

TABLE 5.1. Recent Community Engagement Projects by the Lyle Center for Regenerative Studies at California State Polytechnic University, Pomona

Community	Activity	Economic Hardship Index (1–100)
El Monte	Participatory community appraisal of neighborhood issues and opportunities	75.9
Pomona	Sustainability workshops for children and families; community visioning exercises and scenario thinking regarding healthy environments	67.4
Watts (South Los Angeles)	Community food assessment	73.3
Tijuana, B.C., Mexico	Community mapping and appraisal	NA

NOTE: Lower hardship index scores reflect stronger economic conditions; higher scores indicate greater economic hardship. Compare the scores to the Los Angeles County Median Index score of 52.5 (County of Los Angeles 2010).

State Polytechnic University, Pomona, has focused on advancing environmental sustainability by engaging many such communities. The center has recently launched its Regenerative Communities Initiative, an effort to focus its outreach on communities of hardship. The center's approach emphasizes the development of community support systems such as food, water, and energy that regenerate themselves over time, ensuring that critical resources will be available for future generations. Table 5.1 lists selected communities where the center has partnered recently, along with their Los Angeles County economic hardship indices. In addition to high rates of economic hardship, these communities suffer from the presence of environmental negatives such as poor air quality and soil contamination as well as the absence of environmental positives such as open space for recreation and healthy, nutritious food.

Why do academic institutions get involved in these types of engagement activities with marginalized communities? While motivations for student involvement in community service have been well studied, motivations of faculty and administrators who typically organize these activities are less understood (O'Meara 2008). KerryAnn O'Meara identifies seven types of motivation for faculty involvement in community service. Most noteworthy are: (1) the facilitation of student learning; (2) achievement of disciplinary goals; (3) personal commitment to specific social issues, people, or places; and (4) personal or professional identity. Indeed, these motivations seem to be reinforced by anecdotal observations from the Lyle Center's involvement in such projects over recent years. The adverse environmental conditions and associated justice issues faced by these communities provide rich and compelling case studies for students who are learning principles of regeneration designed to advance sustainability; Lyle Center participants believe they possess technical knowledge which may be helpful in improving conditions faced by these communities. In addition, many faculty and students believe they possess a professional responsibility to address these conditions and injustices, and many simply see a community in need and wish to help alleviate their suffering.

Undoubtedly, academic partners bring much-needed expertise, energy, and resources to these engagement activities. But what are the unintended consequences of academic institutions helping communities in need? Do these consequences promote or inhibit the advancement of sustainable communities? This chapter examines these questions through a review of critiques on community engagement processes from education, social service, and community-organizing literature and offers the concept of ecological sovereignty as a frame for considering partnerships between academic institutions and local communities.

Critiques of Outside Assistance and the Potential of True Generosity

While well intentioned, community service by academic institutions is often vulnerable to claims of exclusion, ignorance of power structures, and preserving the interests of the institution at the expense of community empowerment. In her study of university community service projects focused on food resources, Julie Guthman (2008) found that students often imposed their own cultural constructs regarding desirable food on partner communities, resulting in mutual frustration and low success rates for implementation. Student participants often failed to fully appreciate racial differences between themselves and community members, sometimes resulting in unintended exclusionary practices through the promotion of "whitened cultural histories" (434).

In addition to exclusion, university service participants may also ignore existing power structures and societal relationships. The desire to help address an immediate need within the community may take precedence over a desire to address underlying conditions contributing to the need. As Katherine Crewe and Ann Forsyth (2003) note in their typology of plural design approaches in landscape architecture, the community service practitioner often consciously chooses to ignore or discount larger social injustices for the sake of dealing with an immediate need in the community. This choice may result from a sense of inefficacy in dealing with larger injustices or a lack of awareness concerning the relationship between the local issue and the societal problem. Perceived disciplinary boundaries may also contribute to this ignorance, if it is believed that the larger injustice is outside the scope of the practitioner's expertise or obligations. While many service approaches may be successful in accomplishing short-term or small-scale benefits for the community, the failure to address problematic power structures may not be conducive to long-term or sustainable solutions.

Although the maintenance of existing power structures and societal relationships may not be a conscious objective, some would argue that the preservation of these structures clearly serves the interest of community service partners as opposed to the local community. In his study of social service organizations, John McKnight described the condition of "Clienthood" (1996, 51) in which the professionalization of services transforms citizens into clients dependent upon relationships with the organizations that serve them, and then are defined wholly by their needs as opposed to their abilities or assets. Viewed in this way, the perpetuation of those needs is essential, in order to preserve the need for ongoing professional work.

Educational theorist Paulo Freire characterized this desire among those in power to simultaneously meet needs while perpetuating them as false generosity, stating: "In order to have the continued opportunity to express their 'generosity,' the oppressors must perpetuate injustices as well" (1970, 45). The perpetuation of injustices through false generosity or McKnight's clienthood concept may be further reinforced through methods of community engagement. Jeffrey Juarez and Kyle D. Brown (2008) offer a framework for understanding the range of community engagement practices in landscape architecture. They characterize extractive community practice as eliciting information from communities for use by outside experts to solve community problems. The landscape architect as outside expert is the principal investigator charged with defining problems, collecting data, and generating plans and solutions. Information obtained through these extractive processes is used, owned, and analyzed by the outside expert. This needs-based approach assumes that the outside expert and her expertise is the answer to the community problems, and that the only role the community can effectively play is that of informant. The result, as Paul Schmitz recently described, is that "Citizens learn to believe that they cannot know whether they have a need, cannot know what the remedy is, cannot understand the process that purports to meet or remedy the need, and cannot even know whether the need is met unless professionals express their own satisfaction" (2011, 115).

Many involved in community outreach recognize the pitfalls of this extreme extractive approach and embrace many practices that tend toward an empowering approach characterized by Juarez and Brown. Moreover, critiques of extractive practices do not suggest that the desire to help is malicious, merely self-serving, or otherwise inappropriate as a professional or personal goal. However, they raise important questions about the often-unintended consequences of our actions, and prompt service providers to reflect critically about their intentions and methods when working with marginalized communities.

Some of these communities have recognized the problems associated with outside organizations seeking to help. Many who work regularly with such communities hear expressions of being "studied to death" without meaningful change. Others may lament the inability or unwillingness of local governments and relief organizations to implement change. Speaking to Lyle Center representatives about their experience

with a project in the Watts community, one community member stated, "We cannot rely on the city to provide any improvements we desire. If things are going to change, then we need to change them for ourselves, and not wait for the city."[1] Such sentiments may reflect the inefficacy of local government, but they may also reflect a system with a vested interest in maintaining existing power structures and the resulting needs of some communities.

In such situations, community awareness of the futility in seeking help may be heightened, an important step toward self-empowerment, yet their capacity to effect change from within the local community may be quite low. Such conditions suggest a vital role for community service from academic institutions. However, this role should contrast sharply with the outside technical expert that extracts, analyzes, and owns information about the community in formulating recommendations. A preferred process focuses on empowering local communities to plan and act for themselves— in other words, to build community capacity. This process rejects the false generosity and dependency of "clienthood" in favor of a process that moves towards what Freire described as true generosity, which "lies in striving so that these hands—whether of individuals or entire peoples—need to be extended less and less in supplication, so that more and more they become human hands which work, and working, transform the world" (1970, 45).

The Concept of Sovereignty

The empowerment of local communities can be described as advancing a form of sovereignty, a term which is gaining favor, particularly in the international development literature concerning food systems. Food sovereignty has been defined as the right of nations and people to control their own food systems, including their own markets, production modes, food cultures, and environments (Wittman, Desmarais, and Wiebe 2010). More recently, the concept has been applied to domestic communities within the United States. Communities such as Sedgwick, Maine, have begun to adopt food sovereignty ordinances aimed at asserting greater control over local food resources and decisions that have an impact on food traditions (Huff 2011). While implying self-sustaining food strategies for communities, the concept is not meant to suggest isolationist or protectionist practices that inhibit trade or cooperation with other communities. Rather, the emphasis is placed on communities making their own decisions about food that nourishes the community, as opposed to having those decisions made for them via global trade policies, governmental subsidies, multinational corporations, or other external decision-makers.

Food sovereignty is a major goal for many communities seeking to advance environmental, economic, and social sustainability. However, a systems approach to sustainability, as advanced by John Lyle (1994) and others, shows us that, in order to attain sovereignty in one system, a community must also attain sovereignty over other critical systems. In order to gain control over a community food system, a comparable level of control must be exerted over other resource systems essential for food production, distribution, processing, consumption, and waste. Most notably, this includes water, land, and energy resources.

Given the interrelated nature of community systems, a broader goal of ecological sovereignty should be considered, whereby communities exert control over the systems essential for sustainability, including physical systems of food, water, energy, the built environment, and waste, as well as relevant social/cultural systems and local ecological systems. Cast in this light, local struggles concerning the presence of environmental negatives or the lack of environmental positives become important discussions about how decisions are made

Fig. 5.1. Pomona residents discuss urban wildlife sightings in their neighborhood at a community event. Photograph by author.

and in what way they promote or detract from local sovereignty or perpetuate injustices (fig. 5.1).

Pathways to Ecological Sovereignty

If marginalized communities accept the notion that ecological sovereignty is a goal worth considering in terms of advancing sustainability, how is it achieved? What is the role of the academic partner? In contrast to efforts that maintain existing social structures, an approach based on sovereignty requires the development of pathways that are transformative for both the communities and those who partner with them. The Regenerative Communities

Initiative at the Lyle Center is focused on the process of nurturing ecological sovereignty in conjunction with partner communities. At the outset, these communities often engage the process with modest, tangible goals such as the establishment of community gardens. As the partner institution, the Lyle Center embraces these goals but also seeks to create conditions where awareness of issues and opportunities is heightened and local knowledge is applied to the transformation of all systems essential for sustainability over multiple generations.

LOCAL SOCIAL CONSCIOUSNESS

There are three such pathways to ecological sovereignty, the first of which is the cultivation of critical social consciousness within local communities. Such consciousness focuses on a clear recognition by community participants of the environmental, economic, and social structures that are shaping conditions in their community. This may lead to a better understanding of factors such as those leading to decisions to locate undesirable land uses within a community, or why fresh produce may be so difficult to obtain from markets within the neighborhood. Educational theorist Ira Shor describes critical social consciousness as making "broad connections between individual experience and social issues,

between social problems and the larger social system" (1992, 127–28). Environmental problems must be posed in such a way that they emerge from community experience and "express problematic conditions in daily life that are useful for generating critical discussion" (55). The academic partner cannot identify the problem from an outsider's view; rather, it must be revealed and deemed relevant to daily life by community members. This approach to community education and empowerment supports informed action or agency by the participant, and contrasts strongly with more conventional forms of education that Freire described as "banking" (1970, 72), where students are essentially seen as empty vessels waiting for deposits of knowledge from teachers. This approach is more closely aligned with "real education" as described by well-known community organizer Saul Alinsky: "Real education is the means by which the membership will begin to make sense out of their relationship as individuals to the organization and to the world they live in, so that they can make informed and intelligent judgments" (1971, 124).

Landscape architecture scholars Kyle D. Brown and Todd Jennings (2003) argue that the development of critical social consciousness within practitioners themselves is just as important as fostering this consciousness within local communities.

Fig. 5.2. Elementary students and their parents explore biofuels and their potential as a community energy source at a Lyle Center workshop, California Polytechnic University at Pomona. Photograph by author.

Only then can practitioners be fully cognizant of the consequences of their professional action—whether it serves to maintain existing social structures or transform them in response to justice or some other motivation. Building on the concepts of Shor, this new mindfulness offers a socially conscious approach to design studio education for practitioners, which promotes critical social consciousness in the context of community service-learning.

The development of critical social consciousness is grounded in an empowering approach to working with communities as characterized by Juarez and Brown and

is the focus of outreach efforts through the initiative. This work has included participatory mapping with mothers in informal settlements in Tijuana, Mexico (Brown and Kjer 2007), community engagement activities in El Monte, California (Juarez and Brown 2008), and more recently, sustainability workshops and environmental visioning activities with schools and community groups in Pomona, California (fig. 5.2). A central aim of these activities is to facilitate the local awareness of environmental, economic, and social structures shaping daily life and opportunity in these communities, and empowering community members to identify and act upon strategies that may improve local conditions.

ASSET-BASED PERSPECTIVES

A second pathway to ecological sovereignty promotes an asset-based perspective in community culture. Often marginalized communities are first and foremost defined by their deficits or characteristics perceived (both by internal and outside observers) as poor in quality or altogether absent. The often-used term "communities in need" embodies this deficit perspective. This deficit mindset is often strongest among community residents and local decision-makers who begin to believe in the inferiority of their community, and as such cannot envision alternative futures where local conditions improve, particularly without significant outside investment or assistance. Emergent practices described as "asset-based community development" (Mathie and Cunningham 2003) tend to emphasize community planning and action around strengths. These practices emphasize a regenerative approach to community development and promote growth based on internal assets as opposed to relying on external resources or investment. This optimistic perspective enables alternative visions of community improvement where

outside investment has not merely displaced the deficit-challenged community of today. This transformation from a deficit culture to an asset culture is a challenging but necessary step toward achieving ecological sovereignty because it creates conditions for empowerment over local decision-making.

COMMUNITY LEADERSHIP

A third pathway cultivates leadership from within the community. Achieving this has been identified as a challenge for the sustainability of local action due to expectations of professionals leading the process and community members lacking ownership of the process (Juarez and Brown 2008). However, recent community-organizing efforts have emphasized the multiplication and sharing of leadership from within and noted that local people are often best positioned to understand local assets, including the recognition of those among them with leadership potential. This perspective contrasts strongly with what some have described as the "trap of heroic leadership." According to Schmitz (2011), practitioners subject to this trap assume they, as the outside expert, know best, that their own cultural values are better than those of the community they are serving, that those communities are defined only by their needs, that simple solutions to complex problems are often adequate, and that cultural differences can or should be ignored. By contrast, an approach that multiplies and shares leadership is one in which we "replace notions of 'helping others' or 'fixing others' with a sense of being in community with others" (Komives and Wagner 2009, 31). This approach necessitates conscious efforts to identify, nurture, and support those within the community with strong capacity for leadership. As Schmitz and Sullivan note, today's leaders "build bridges, establish free spaces where citizens can be supported as community change agents and problem solvers, and continuously foster the emergence of new leaders" (1997, n.p.).

Emerging Questions about Ecological Sovereignty

As we consider the importance and relevance of ecological sovereignty to the practices of community service by academic institutions, a number of questions emerge which may guide future inquiry. A primary question concerns the limits of sovereignty as a concept. Where does sovereignty end and protectionism or exclusionary practice begin? As opposed to isolating the community from its neighbors, the intention of this concept is to promote local choice over systems and resources that serve a marginalized

community. However, any quick Internet search of the word "sovereignty" reveals significant usage of the term in North America to refer to anti-authoritarian movements of "sovereign citizens" aimed at challenging local, state, and federal law, as well as the role of international cooperation. Indeed, taken to its extreme, ecological sovereignty could serve as a rallying cry for those seeking to secure resources for themselves, separate themselves from a wide variety of societal influences, or withdraw into exclusionary practices based on race, faith, or other cultural differences. The potential for this concept to be used to further a variety of ends thus signals its potency, and raises important questions about its role in community development.

A second question relates to the appropriateness of advocacy for societal transformation in community service by academic institutions. Is the transformation of power relations an appropriate goal for community engagement work? If so, it raises sharp questions about the political position of those in higher education who participate in this engagement. No longer would academics be able to hide behind the veil of objectivity, professional neutrality, and technical assistance. While many argue that objectivity is already a myth, it remains a powerful identifying construct for much mainstream academic work. As scholars struggle with the idea of "finding center" in landscape architecture and related disciplines, they must consider the larger issue of our role in society along with the appropriateness of activism in the professional and academic realms. To this end, a closer examination of the relationship between academic and professional practice may be warranted. While many call for continued and strengthening bonds between these realms, the allegiance with professions rooted in existing power structures may inhibit opportunities to explore alternative futures, particularly ones which advance justice for marginalized groups.

A final question relates to the implications of ecological sovereignty for the disciplines engaged in community service related to sustainability. These disciplines include a diverse set of fields such as geography and other social sciences, cultural and landscape studies, engineering, and the disciplines of environmental design. If we accept the premise that sustainability activism is a worthy pursuit for these disciplines and that sovereignty is a necessary component for sustainability in marginalized communities, how should the discourse and curricula in these fields change? Certainly the development of service-learning pathways related to critical social consciousness, asset-based development, and multiplication of leadership suggests new skills and abilities

in community organizing, facilitation, and community education that are not presently emphasized in professional/technical education and its various accreditation guidelines. In addition to practical skills and experience in working effectively with communities, some academic programs may also need greater depths of theoretical inquiry related to the motivations, roles, and responsibilities of the community-service professional. Finally, in service-learning projects, student participants must be supported in the awareness and development of their individual positions within this work. This will mean a shifting of the "center" of all academic disciplines concerned with the contemporary landscape, and perhaps a broadening or even poly-centered restructuring of pedagogy for such disciplines in coming years.

NOTES

1. During outreach activities, the author received this comment from a person who wishes to remain anonymous.

REFERENCES

Alinsky, Saul. 1971. *Rules for Radicals.* New York: Vintage Books.

Brown, Kyle D., and Todd Jennings. 2003. "Social Consciousness in Landscape Architecture Education: Toward a Conceptual Framework." *Landscape Journal* 22 (Fall): 99–112.

Brown, Kyle D., and Tori Kjer. 2007. "Critical Awareness in the Era of Globalization: Lessons for Landscape Architecture from an Informal Community in Tijuana, Mexico." *Landscape Review* 12: 26–45.

County of Los Angeles. 2010. "Life Expectancy in Los Angeles County." Department of Public Health report.

Crewe, Katherine, and Ann Forsyth. 2003. "Landscapes: A Typology of Approaches to Landscape Architecture." *Landscape Journal* 22 (Spring): 37–53.

Freire, Paulo. 1970. *Pedagogy of the Oppressed.* New York: Continuum.

Guthman, Julie. 2008. "Bringing Good Food to Others: Investigating the Subjects of Alternative Food Practice." *Cultural Geographies* 15:431–47.

Huff, Ethan A. 2011. "Maine Town Becomes First to Declare Food Sovereignty." *NaturalNews.* www.naturalnews.com/031667_food_freedom_Maine.html (accessed November 10, 2014).

Juarez, Jeffrey, and Kyle D. Brown. 2008. "Extracting or Empowering? A Critique of Participatory Methods for Marginalized Populations." *Landscape Journal* 27 (Fall): 190–204.

Komives, Susan, and Wendy Wagner. 2009. *Leadership for a Better World.* San Francisco: Jossey-Bass.

Lyle, John T. 1994. *Regenerative Design for Sustainable Development.* New York: Wiley.

Mathie, Alison, and Gord Cunningham. 2003. "From Clients to Citizens: Asset-Based Community Development as a Strategy for Community-Driven Development." *Development in Practice* 13:474–86.

McKnight, John. 1996. *The Careless Society: Community and Its Counterfeits.* New York: Basic Books.

O'Meara, KerryAnn. 2008. "Motivation for Faculty Community Engagement: Learning from Exemplars." *Journal of Higher Education Outreach and Engagement* 12 (1): 7–29.

Schmitz, Paul. 2011. *Everyone Leads: Building Leadership from the Community Up.* San Francisco: Jossey-Bass.

Schmitz, Paul, and L. Sullivan. 1997. "Practicing What We Preach: Creating Transforming Organizations." *Wingspread Journal* 19 (4): n.p.

Shor, Ira. 1992. *Empowering Education: Critical Teaching for Social Change.* Chicago: University of Chicago Press.

Wittman, Hannah, Annette Desmarais, and Nette Wiebe. 2010. *Food Sovereignty: Reconnecting Food, Nature and Community.* Oakland, CA: Food First Books.

MARTIN J. HOLLAND

Memory Work

THE SUBMISSIONS TO THE OKLAHOMA CITY
MEMORIAL COMPETITION

On the morning of April 19, 1995, Timothy McVeigh parked a Ryder rental truck containing a homemade 4,800-pound explosive concoction of fertilizer and diesel fuel at the northern entrance of the Alfred P. Murrah Federal Building in downtown Oklahoma City. The blast that resulted from the detonation of this bomb shattered windows in a ten-block radius, caused severe structural damage to four adjacent buildings, and prompted the northern section of the Murrah Building to collapse, killing 168 people, 19 of whom were children. Just four days after the bombing, while search-and-rescue operations were still being conducted, the *Daily Oklahoman*

(the newspaper of record for Oklahoma City) ran an editorial that concluded with these words: "For them [the victims], perhaps Oklahomans should build a memorial on the very site of this tragedy and list each of their names. It might be topped with a bell that tolls each day at 9 a.m.—a place to bring wreaths and flowers, notes and small toys for the children, a place to weep. Let us remember our dead, not as a group, but as individuals whose lives had meaning, who contributed to this city. Let us remember them more than we remember the tragedy itself. Let us honor our dead with dignity. As we always have" (1995, 6).

This public call for a memorial resonated

with a traumatized populace, and elected officials were soon inundated with letters, faxes, and phone calls supporting a commemorative effort. The speed of local, state, and federal responses to this public demand was staggering. In the space of less than a year, Oklahoma City mayor Ron Norick formed the Oklahoma City Memorial Task Force (MTF), which organized and held an international memorial competition that received 624 entries from twenty-three countries. In July 1997, the Berlin-based Butzer Design Partnership, formed by Hans and Torrey Butzer with their associate Sven Berg, was selected as the winners of the competition. Three years later, President Bill Clinton officially dedicated the memorial on the fifth anniversary of the bombing.

The design submissions and the memorial process in Oklahoma City are worthy of our attention as designers of the built environment. The chair of the Memorial Task Force, real estate attorney Robert M. Johnson, lauded Oklahoma's memorial process, stating to a congressional subcommittee that "We *democratized* the memorial process by making it open and inclusive. There have been no political, socio-economic, or other barriers to participation. Most importantly, we have encouraged, solicited, and given great deference to participation by family members and survivors in all aspects of this memorial

process" (U.S. Congress 1997, 47; emphasis added). Edward T. Linenthal, professor of history at Indiana University and author of *The Unfinished Bombing: Oklahoma City in American Memory*, also praised the work of the MTF as "a significant event in the history of public memorialization," noting that "it would be hard to find another memorial process so consciously designed to be *therapeutic*, to help individuals become a bereaved community, and to engage mass murder through memorial planning" (2001, 229; emphasis added).

The submissions are examples of what James E. Young, professor of English and Judaic studies at the University of Massachusetts–Amherst and jury member of the World Trade Center Site Memorial Competition, has called *memory work*, the intentional labor that individuals have undertaken to mark or recognize a particular event (Young 2006, 214). However, certain aspects of the Oklahoma process, especially its compressed time frame, yielded a very particular kind of memorial. In this light, Johnson's and Linenthal's congratulatory claims beg for critical interrogation. The process was democratic in the sense that "the people"—a still-traumatized people—were directly involved in an expedited memorial process, producing an inherently therapeutic commemorative space. Critical scrutiny of the "memory work" of the Oklahoma City

Memorial reveals commemorative gestures that disassociate, distract, and distance memorialization from the actual, tragic history of the site. Having examined the entire body of the Oklahoma Memorial competition's submissions, which includes work by professional design firms as well as amateurs, in this chapter I document certain recurring themes that are representative of larger values at work within the competition, and within American commemorative culture at the end of the twentieth century.

Recurring Themes

VERNACULAR SENTIMENTALITY

From 624 official entries to the Oklahoma City International Memorial Competition, several distinct themes emerge. One of the most common expressions of public memory took the form of *vernacular sentimentality,* containing expressions of popular material culture such as teddy bears and other stuffed animals, symbolic ribbons, colorful rainbows, stylized "broken" hearts, and winged angels—what some refer to as "kitsch" (Sturken 2007). One design entry even included all of these elements— angels, a broken heart on a memorial wall, a children's memorial with nine teddy bears, and a multistoried commemorative

ribbon (fig. 6.1). These motifs are considered inherently comforting; however they also often unwittingly infantilize the victims and reduce complicated and emotionally trying circumstances to commodified expressions of sadness, comfort, or hope. The act of leaving tokens of sympathy in the immediate aftermath at a site of a tragic event is a common social practice; in turn, the "translation" of such left-behind objects into a larger commemorative site design strategy is unsurprising.

The preoccupation with the loss of children dominated the coverage of the Oklahoma City bombing to such an extent that a frustrated and grieving relative described his lost twenty-nine-year-old step- daughter in these terms: "Like the children, she had no cause, no politics, no enemies and no chance. I am one of many who mourn the loss of a child" (McMullen 1995, 90A). Preoccupation with the young children was exemplified by the image of fireman Chris Fields holding the lifeless body of the infant Baylee Almon, which became an iconic image in the days after the bombing, framing the bombing as an attack on those most innocent and vulnerable. Families who lost a child were placed near the top of a "memorial hierarchy" and in turn garnered widespread sympathy and support. Indeed, the only families that had more social capital

Fig. 6.1. Sentimental themes. Photograph courtesy of the Oklahoma City National Memorial and Museum, Coll. No. 1592.

were those unfortunate enough to have lost multiple family members in the attack (Linenthal 2001, 197). The reliance on vernacular sentimentality is representative of a larger trend in American public commemoration, which favors emotions and feelings over the critical engagement of the site's history, ignoring what happened and why (Doss 2010).

PATRIOTISM

In the immediate aftermath, coverage of the bombing emphasized that it had taken place in "America's Heartland," constructing Oklahoma City as paradigmatic of the nation itself. The initial news reports accused Islamic terrorists as being responsible, but it soon became clear that the perpetrator

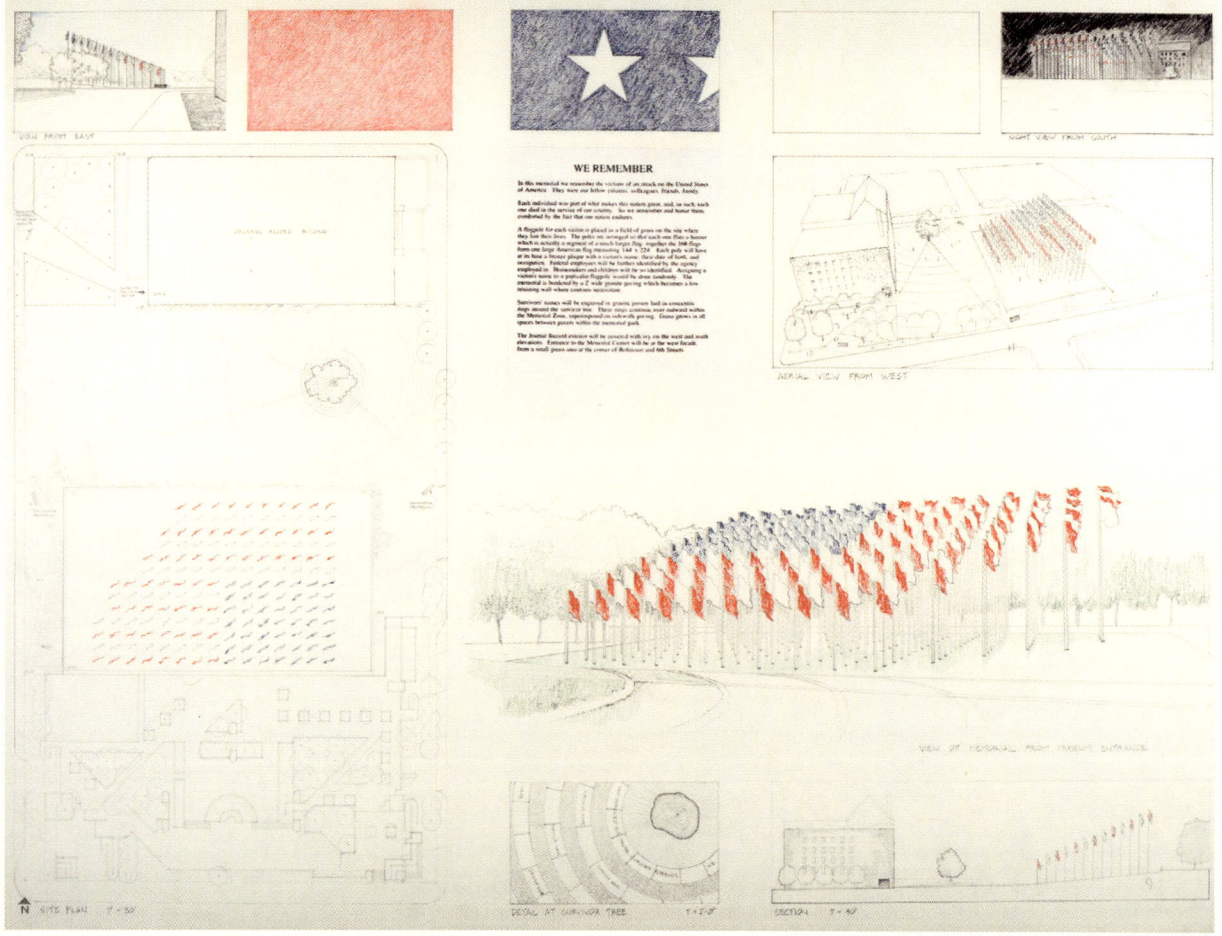

Fig. 6.2. Patriotic themes. Photograph courtesy of the Oklahoma City National Memorial and Museum, Coll. No. 1754.

was a homegrown terrorist, expressing what he believed to be his own form of civic duty. Since the target of the bombing was a U.S. federal building, many equated the attack with a military strike against America, resulting in the extensive use of patriotic imagery in many of the memorial submissions. The initial anxiety that was produced by this act of domestic terrorism was dispelled through the extensive use of national symbols, indicating that the country would persevere and overcome this tragedy. Designs included such iconography as the American flag (with its red, white, and blue color scheme), the use of stars and stripes, bald eagles, and even the outline of the nation with its contiguous forty-eight states.

In one submission (fig. 6.2), the placement of 168 flagpoles of increasing height, each adorned with a colorful banner, works to form a large, albeit unfinished, *Stars and Stripes*. It equates the "torn" fabric of the flag with the wounding that was experienced by the citizens of the city and of the nation. Although 168 individuals are memorialized, each is reduced to the identity of "American," and visitors to the memorial are invited to share in that sense of a communal, nation-centered identity.

The language normally reserved for those in military service framed the discussion concerning those lost in Oklahoma City. During the dedication service of the memorial grounds on April 19, 2000, President Clinton stated, "there are places in our national landscape so scarred by freedom's sacrifice that they shape forever the soul of America—Valley Forge, Gettysburg, Selma. This place is such sacred ground." Linenthal shares that these words "grated" on him, as "these people's lives were not given in an act of conscious sacrifice for their nation; they were taken in an act of mass murder" (2001, 234). The conflation of normal acts of daily life, such as going to work or school or even seeing a movie, with heroic acts of sacrifice reifies the nation and her people. There were certainly hundreds of heroic acts performed that day, some by the injured themselves, but to equate the deaths of people going about their everyday lives with bravery in combat diminishes those who consciously sacrifice their lives for their country, and

unwittingly lessens the true horror of mass murder. Unlike the 9/11 attacks that occurred six years later, this was an act inflicted upon the country by one of its own citizens—an Eagle Scout and a decorated member of the American armed forces. To rely on national symbols which promote and assert the continued existence of the nation-state distorts and distracts from the realities of this particular tragedy.

RELIGIOUS ICONOGRAPHY

Christian iconography as a form of religious faith was another recurring theme of the submissions. Some entries contained elements such as statuary of angels interceding or assisting the victims of the bombing. Others included expressions of faith such as people kneeling in prayer, statuary of large praying hands, or the presence of religious texts (such as the Ten Commandments) and biblical scriptures etched onto various surfaces.

In one submission (fig. 6.3), the proposed memorial plan takes the form of an expansive outdoor cathedral, open and exposed to the elements with superfluous flying buttresses. The use of traditional cathedral architecture, including the presence of the nave, transept, apse, and ambulatory, creates a cruciform ground plan that carves a literal cross in the landscape. The designer also provocatively

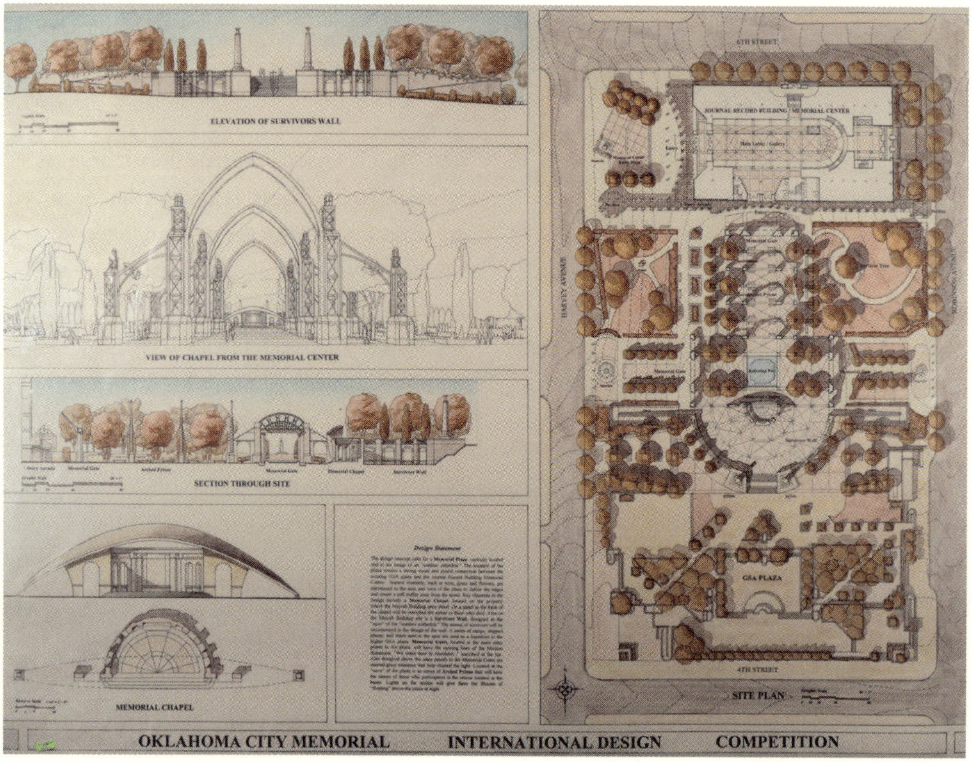

Fig. 6.3. Religious themes. Photograph courtesy of the Oklahoma City National Memorial and Museum, Coll. No. 1388.

places the specific point of detonation at the location traditionally reserved for the altar. This placement asserts the notion of sacrifice, not in terms of national or military service but rather within a religious context. Suffering, in this case the "sacrifice" of the bombing's victims, is given meaning and suggests two possible religious narratives. The first concerns the ongoing battle of good versus evil, where the presence of the cross implies that good will triumph over evil. The second suggests the redemptive function of suffering, that suffering is necessary for salvation from the pervasiveness of human sin. While assertions of religious beliefs may be inappropriate for a civic, public memorial, it is not surprising that Americans default to a religious vocabulary especially when these religious narratives help provide an explanation and a deeper meaning at a point of real human suffering.

NATURE

Throughout the memorial competition, the natural world was frequently referred to as a restorative agent. Nature, specifically in the form of a garden, possesses cultural associations with the idea of an unobtainable paradise—for example, the Garden of

Eden, mythic Arcadia, and even the state of Utopia. Numerous submissions depicted the site as a vast and intricate garden, often to such a degree that the surrounding urban context and cityscape were simply ignored. As Kenneth Helphand observes in *Defiant Gardens,* "gardens promise . . . hope over despair . . . life in the face of death." Gardening in environments that are usually associated with pain, suffering, and inhumanity (for example, prison camps,

military front lines) is an act of defiant optimism (2006, 7).

In one example (fig. 6.4), the design not only depicts the site as a lush and colorful garden, but also relies on the care and nurturing involved with plant propagation to commemorate the tragedy. Using a series of curved glass walls offset from the foundation of the Government Service Agency plaza, on a slight angle of the footprint of the former Murrah Building, this entry draws a parallel

Fig. 6.4. Themes of nature. Photograph courtesy of the Oklahoma City National Memorial and Museum, Coll. No. 1315.

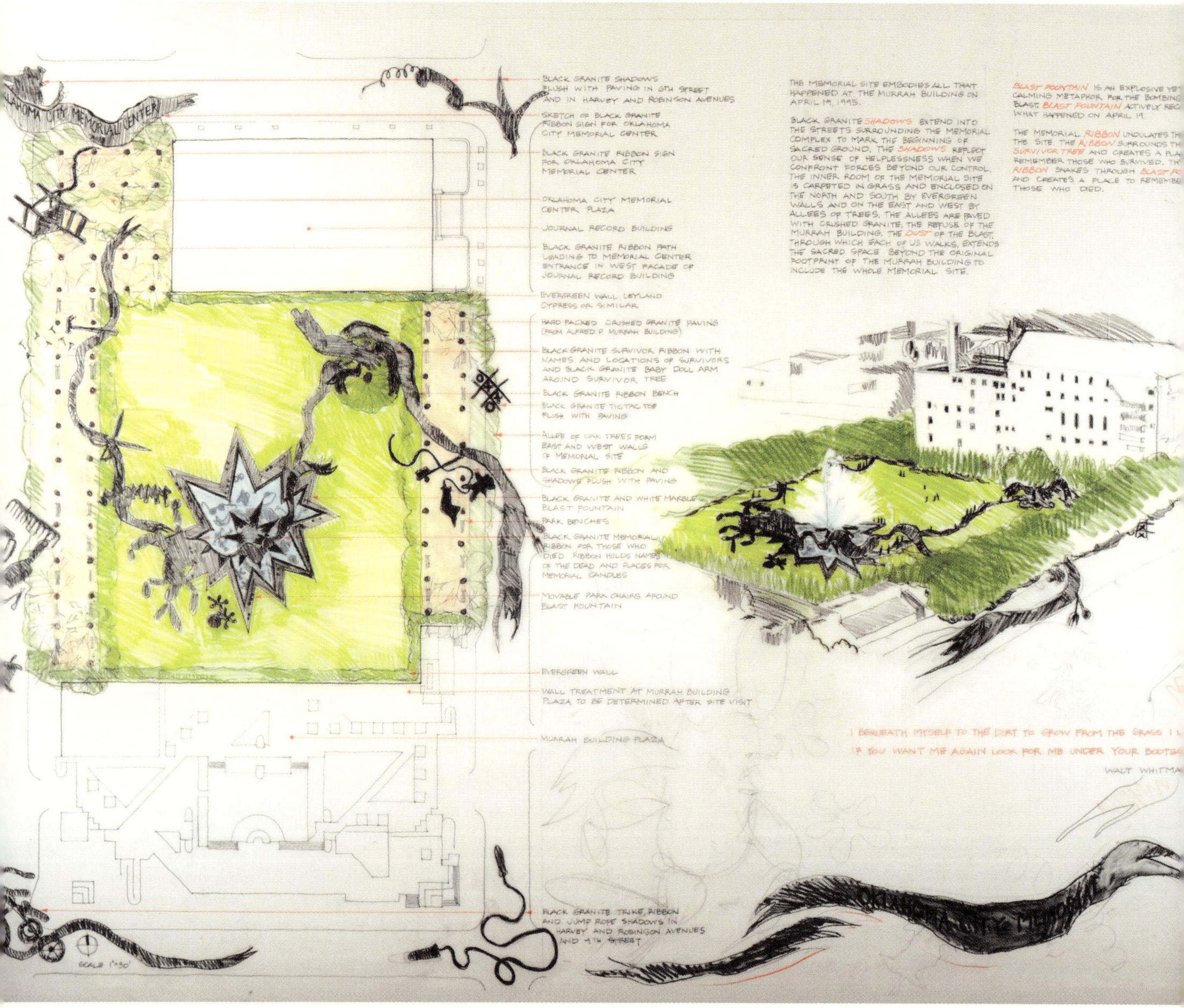
OKLAHOMA CITY MEMORIAL CENTER
BLACK GRANITE SHADOWS FLUSH WITH PAVING IN 6TH STREET AND IN HARVEY AND ROBINSON AVENUES
SKETCH OF BLACK GRANITE RIBBON SIGN FOR OKLAHOMA CITY MEMORIAL CENTER
BLACK GRANITE RIBBON SIGN FOR OKLAHOMA CITY MEMORIAL CENTER
OKLAHOMA CITY MEMORIAL CENTER PLAZA
JOURNAL RECORD BUILDING
BLACK GRANITE RIBBON PATH LEADING TO MEMORIAL CENTER ENTRANCE IN WEST FACADE OF JOURNAL RECORD BUILDING
EVERGREEN WALL, LEYLAND CYPRESS OR SIMILAR
HARD PACKED CRUSHED GRANITE PAVING (FROM ALFRED P. MURRAH BUILDING)
BLACK GRANITE SURVIVOR RIBBON WITH NAMES AND LOCATIONS OF SURVIVORS AND BLACK GRANITE BABY DOLL ARM AROUND SURVIVOR TREE
BLACK GRANITE RIBBON BENCH
BLACK GRANITE TIC TAC TOE FLUSH WITH PAVING
ALLEE OF OAK TREES FORM EAST AND WEST WALLS OF MEMORIAL SITE
BLACK GRANITE RIBBON AND SHADOWS FLUSH WITH PAVING
BLACK GRANITE AND WHITE MARBLE BLAST FOUNTAIN
PARK BENCHES
BLACK GRANITE MEMORIAL RIBBON FOR THOSE WHO DIED. RIBBON HOLDS NAMES OF THE DEAD AND PLACES FOR MEMORIAL CANDLES
MOVABLE PARK CHAIRS AROUND BLAST FOUNTAIN
EVERGREEN WALL
WALL TREATMENT AT MURRAH BUILDING PLAZA TO BE DETERMINED AFTER SITE VISIT
MURRAH BUILDING PLAZA
BLACK GRANITE TRIKE, RIBBON AND JUMP ROPE SHADOWS IN HARVEY AND ROBINSON AVENUES AND 4TH STREET
SCALE 1"=30'
THE MEMORIAL SITE EMBODIES ALL THAT HAPPENED AT THE MURRAH BUILDING ON APRIL 19, 1995.
BLACK GRANITE SHADOWS EXTEND INTO THE STREETS SURROUNDING THE MEMORIAL COMPLEX TO MARK THE BEGINNING OF SACRED GROUND. THE SHADOWS REFLECT OUR SENSE OF HELPLESSNESS WHEN WE CONFRONT FORCES BEYOND OUR CONTROL. THE INNER ROOM OF THE MEMORIAL SITE IS CARPETED IN GRASS AND ENCLOSED ON THE NORTH AND SOUTH BY EVERGREEN WALLS AND ON THE EAST AND WEST BY ALLEES OF TREES. THE ALLEES ARE PAVED WITH CRUSHED GRANITE, THE REFUSE OF THE MURRAH BUILDING, THE DUST OF THE BLAST, THROUGH WHICH EACH OF US WALKS, EXTENDS THE SACRED SPACE BEYOND THE ORIGINAL FOOTPRINT OF THE MURRAH BUILDING TO INCLUDE THE WHOLE MEMORIAL SITE.
BLAST FOUNTAIN IS AN EXPLOSIVE YET CALMING METAPHOR FOR THE BOMBING BLAST. BLAST FOUNTAIN ACTIVELY REC WHAT HAPPENED ON APRIL 19.
THE MEMORIAL RIBBON UNDULATES THROUGH THE SITE. THE RIBBON SURROUNDS THE SURVIVOR TREE AND CREATES A PLACE TO REMEMBER THOSE WHO SURVIVED. THE RIBBON SNAKES THROUGH BLAST FOUNTAIN AND CREATES A PLACE TO REMEMBER THOSE WHO DIED.
I BEQUEATH MYSELF TO THE DIRT TO GROW FROM THE GRASS I LOVE
IF YOU WANT ME AGAIN LOOK FOR ME UNDER YOUR BOOTSOLES
WALT WHITMAN
OKLAHOMA CITY MEMORIAL

Fig. 6.5. Themes of trauma. Photograph courtesy of the Oklahoma City National Memorial and Museum, Coll. No. 1329.

between a greenhouse as a place of care and the commemoration of the dead. The larger landscape is rendered in all of the seasonal colors, indicating occupation throughout the year, and a variety of plant materials of different sizes is also indicated. It is the inherent fragility and yet tenacity of the garden that is particularly appealing and, as one landscape historian notes, "gives privilege to landscape architecture over other forms of memorialization" (Hunt 2001, 22).

TRAUMA

Depictions of trauma or rupture were common within the larger body of official submissions to the Oklahoma City Memorial Competition. Many entries used the point of detonation and the subsequent blast radius of the explosion as an ordering device for the entire site. While one particular submission (fig. 6.5) violates the terms of the competition as it extends past the established site boundaries, the scattering of oversized, everyday objects throughout the downtown core is as chilling as it is confrontational. These sculptural "shadows," as the designer calls them, decentralize the violence from the predictable confines of the site and redistribute it throughout the entire city, radically challenging the notion of a memorial site. This scheme also compels the citizens of Oklahoma City to acknowledge

violence and by disrupting daily routines exemplifies the continual pain that the family members felt after the immediate shock of their losses waned. Yet, given the tyranny of the therapeutic in the memorial process (in Oklahoma City but also in contemporary American commemorative practices in general), these trauma-themed designs would be unlikely candidates for selection.

The Finalists and the Winning Entry

A striking feature of the memorial competition was how similar the five finalists' designs were to one another. Four out of the five provided inherently peaceful and meditative environments in their respective designs. Susan Herrington and Mark Stankard's entry reconceived the site as a new plaza, adjacent to an introduced urban forest, where the site of the former Murrah Building would be transformed into a large set of stairs/seats that would reconnect the Government Services Administration plaza to the existing streetscape. Near the pinnacle of those stairs, a structural glass wall with the names of the 168 victims etched upon it (and containing small mementos that the deceased had valued) would greet visitors.

The entry from Hanno Weber, Kathleen Hess, and Michael Maher conceived of the memorial grounds as an inherently civic

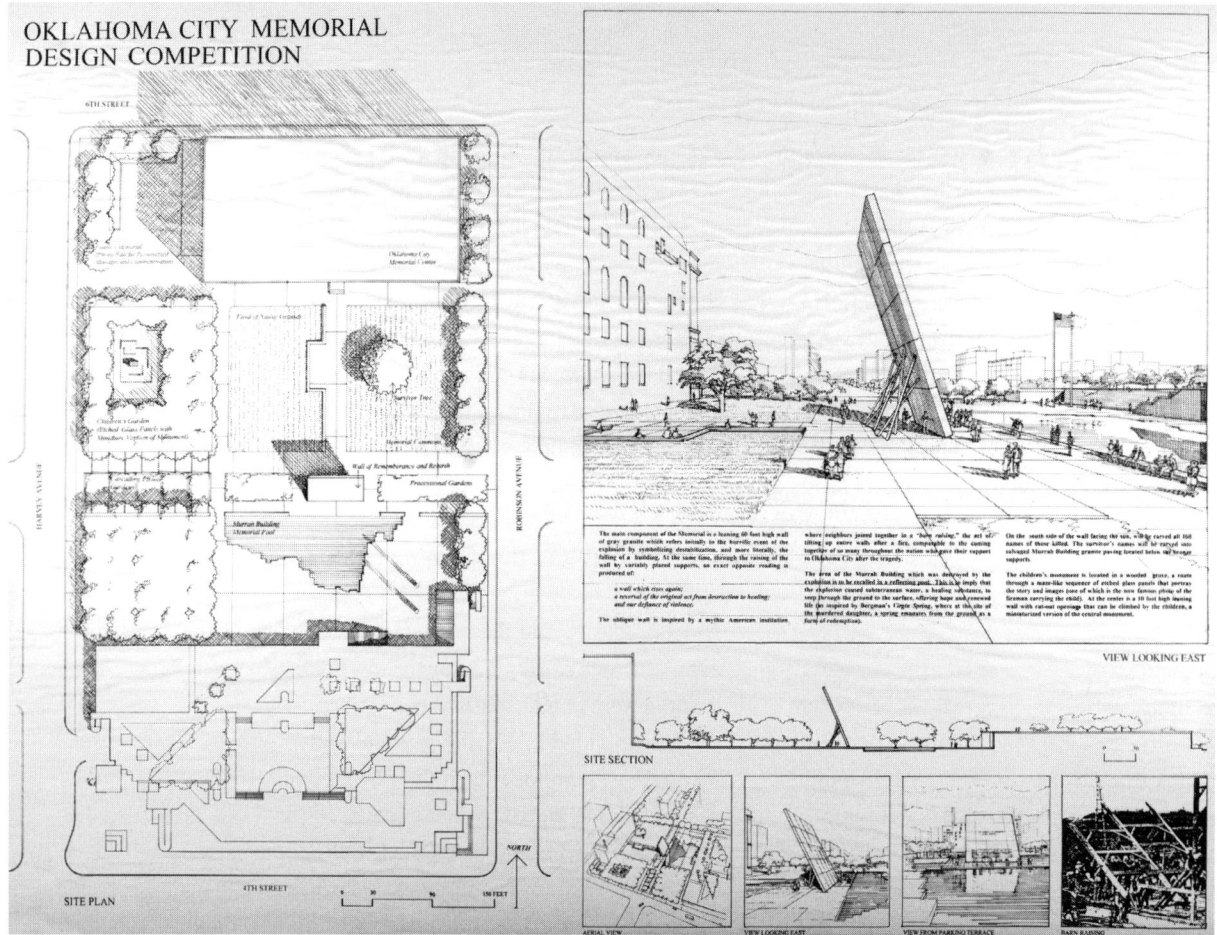

OKLAHOMA CITY MEMORIAL
DESIGN COMPETITION

Fig. 6.6. Rossant and Scherr submission. Photograph courtesy of the Oklahoma City National Memorial and Museum, Coll. No. 2247.

space, but instead of a plaza, they provided a wide, circular, and sloping clearing, surrounded by 168 trees, separated from the rest of the memorial grounds by a thick wall, providing "a peaceful meadow for reflection." Brian Branstetter and J. Kyle Casper offered a linear series of twelve concrete "chapel-like rooms" which contained small apertures to allow the victims' names to be illuminated by sunlight at noon on their respective dates of birth. These three design proposals, as well as the winning design of the Butzers, all succeed in individuating the 168 dead by naming them, a hallmark of contemporary American memorial practices, doing so in a tranquil, therapeutic setting.

The outlier was James Rossant and Richard Scherr's entry (fig. 6.6). Their design's main focus was a sixty-foot-tall leaning granite wall, complete with ad hoc "temporary" buttressing, not only capturing the unease of a wall close to collapsing but also symbolizing an act of communal rebuilding. Their competition board suggests "a wall which rises again. A reversal of the original act from destruction to building: and our defiance of violence" and was inspired "by a mythic American tradition where neighbors joined together in a 'barn raising'" (Rossant and Scherr 1995). It was the ambiguity of this design that ruled it out as a winning entry— "the jury questioned whether or not it would be symbolically understood and whether a memorial should require interpretation" (Collyer 1997, 8). Yet all memorials require a degree of interpretation (unless they are perhaps a ruin), given that memorials are reconstructions and representations of events, acts, and agents that are no longer present. Memorials are points of reference within the larger landscape that serve as aids to cultural and social memory; in effect, they are texts that must be read in order to be understood.

Questioning the necessity of interpretation suggests that the operational mindset of the Design Selection Committee was one that privileged univocality (and simplicity) and was wary of a design that might challenge visitors or require critical engagement with the site and the event. The Design Selection Committee, the final jury charged with selecting the winning entry, comprised seventeen people—nine of whom were victims' family members or were themselves survivors of the bombing. Given the compressed timing of the memorial competition, this still-traumatized community acting as jury insisted upon a memorial that would provide solace and healing.

The winning memorial design by the Butzer Design Partnership (fig. 6.7) consists of seven distinct memorial components, many of them specifically identified with a subgroup of people affected by the bombing. All are located within a larger, tranquil commemorative landscape. The *Survivor's Wall* lists the people who endured the bombing within a four-block radius of the Murrah Building; the *Rescuer's Orchard* "honors the thousands of volunteers that rushed to Oklahoma City in the aftermath of the bombing" (Oklahoma City National Memorial and Museum, 2011). Other memorial components include the *Gates of Time*, a pair of urban-scaled walls that provide major points of egress to the memorial grounds; the *Reflecting Pool*, a taut 318-foot-long by 53-foot-wide shallow pool that reflects the city skyline; the *Survivor's Tree*, a seventy-year-old American elm (*Ulmus Americana*) that was itself severely wounded in the blast and now provides the social center to the memorial grounds; and the *Children's Area* immediately in front of the memorial museum, where large slate "blackboards" are inserted into the paving

Fig. 6.7. Plan of the winning competition entry, Butzer Design Partnership. Reproduced by permission of Butzer Design Partnership.

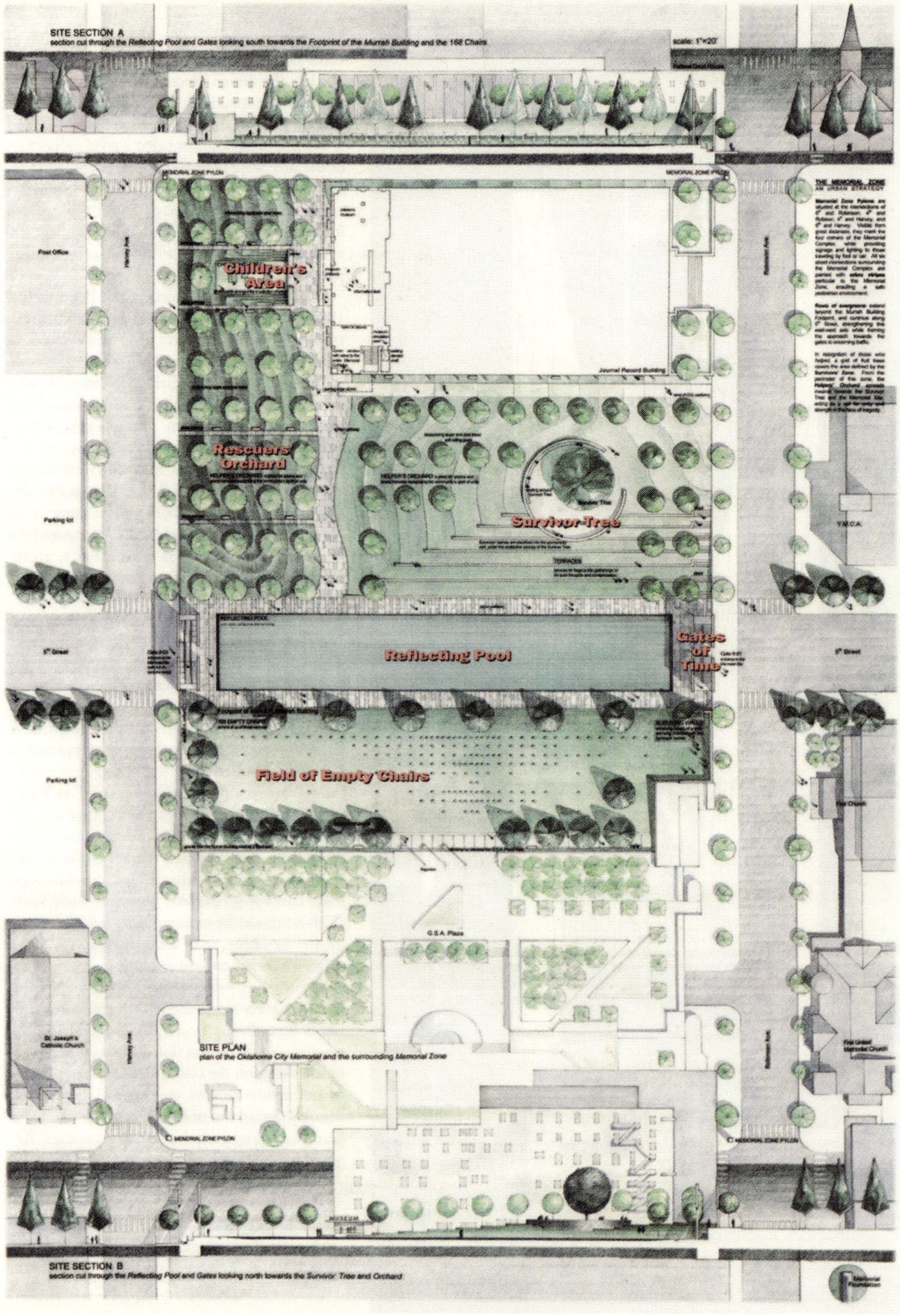

pattern, encouraging children to write and draw their own messages.

It is the *Field of Empty Chairs* that is most commonly associated with the Oklahoma City Memorial. This memorial component consists of a series of stylized glass and bronze chairs, carefully aligned in nine straight rows. A chair honors each victim, with the individual's name etched on the front of the chair's glass base. The location of each memorial chair was carefully placed to signify where each victim would have been at the time of the attack, with the nine rows representing the nine floors of the Murrah Building. This "memorial exactitude" was a critical issue for the designers and the family members, providing the latter with the exact location of where their loved one was during their last moments, and a specific destination to visit and grieve (National Park Service, 2011). Given the unexpected nature of the attack, having a particular place that accounted for the last living moments of a loved one was of paramount importance to the family members of those killed (Linenthal 2001, 195).

Memory and Therapy

Identifying the five design themes— vernacular sentimentality, patriotism, religious faith, nature, and trauma—reveals how professional and amateur designers attempted to represent and commemorate the meaning of the Oklahoma City bombing. Why were the meditative and therapeutic landscapes so popular and even privileged by the Design Selection Committee? I have suggested that the answer lies in part with the composition of the jury as well as the compressed time frame of the competition process. But the answer can also be found in the memorial process that was undertaken to create the guidelines for the memorial competition.

When Robert M. Johnson (U.S. Congress 1997) claimed that the MTF had *democratized* the memorial process, he was referring to the extraordinary steps that he and members of the Memorial Ideas Input Subcommittee performed to solicit public input. To that end, on July 24, 1995, just ninety-one days after the bombing, members attended the first formal gathering of families and survivors. From those discussions the subcommittee established the first working draft of what would become a memorial survey. Table 6.1 reproduces the text of the sample survey form used to solicit public input.

In effect, the survey specified certain responses but also foreclosed a range of other possible responses, and instead offered a piecemeal approach to the memorial which could be justified as simply giving the citizens

TABLE 6.1. Sample Survey Form Developed by the Memorial Ideas Input Subcommittee for the Murrah Federal Building Memorial Task Force

---

MEMORIAL SURVEY

Murrah Federal Building Memorial Task Force

Oklahoma City Mayor Ron Norick formed a volunteer task force of citizens to develop an appropriate memorial regarding the Alfred P. Murrah Federal Building bombing on April 19, 1995. The first step is to find out what we want the memorial to represent. This survey is one of the ways to get your thoughts, feelings and ideas to the Memorial Task Force. Please make copies of this survey and encourage your family, friends and coworkers to share their ideas.

1. What is your zip code? Please write all 5 numbers. _____

2. What is your age? Please check your age bracket.

____ under 12	____ 25–24	____ 55–64
____ 12–17	____ 34–44	____ 65–74
____ 18–24	____ 45–54	____ 75+

3. When you are at the memorial what feeling(s) do you want to have? Please check one or more.

____ Pride	____ Solemn	____ Peaceful
____ Anger	____ Courage	____ Healing
____ Fear	____ Concerned	____ Spiritual
____ Hope	____ Inspired	____ Other _____

4. Please check one or more. The memorial should:

____ Include the names and stories of victims and survivors.
____ Honor those who helped.
____ Not be limited to the Murrah Federal Building site.
____ Include an interpretive center/museum to educate people about the site.
____ Be for the whole nation.
____ Include something for the children.
____ Show the bombing's violence.
____ Describe the community, state and national unity.
____ Have scenes showing before, during and after the bombing.
____ Include a green space with trees and flowers.
____ Show the world-wide response.

Other things the memorial should be or do:

_____

_____

Please mail your completed survey(s) and any additional comments and suggestions as soon as possible and no later than Feb. 15, 1996. Your tax-deductible donation may be mailed to the Murrah Memorial Fund.

POST OFFICE BOX 18390 | OKLAHOMA CITY, OK | 73154-0390 | PHONE: (405) 236-8400

SOURCE: Oklahoma City National Memorial and Museum, 1996.

of Oklahoma City what they wanted as they expressed their desires and wishes. The rapidity of this process, combined with the determinative role of the public survey, was fundamental to the selection of the final, winning design. The Butzers' design provided all of the needed components elicited by the survey.

In many ways, the memorial process in Oklahoma City was an extension of the community outreach that provided immediate triage in the minutes, hours, days, and weeks after the bombing. It was focused primarily on the psychological health of the victims' families. Those who were still in shock understandably struggled with what the event meant for them, or what they would want to remember about the event itself. Robert Johnson spoke of providing "great deference" to those most affected by the trauma. But their influence extended far beyond merely sharing their opinions and feelings about the memorial. They were intimately involved with the entire process, including serving on the final selection jury. The resulting memorial was focused on providing a tranquil and peaceful experience, and each of the survivors, victims, children, and rescuers had a distinct area set aside for them.

As Kirk Savage, professor of history of art and architecture at the University of Pittsburgh, has astutely observed about American memorialization practices, there is a paradox at the heart of the therapeutic monument: "the more intense the focus on each individual victim, the less the monument justifies itself because the less there is to distinguish this particular loss from all other traumatic losses suffered in any society" (2006, 113). The reliance on therapeutic settings and environments forecloses the possibility of other cognitive and emotional responses. While these tranquil and reflective environments might be appropriate for those who are actively grieving and suffering, it is not clear what meaning the memorials offer for others, beyond a place of generic refuge or solace. Moreover, they run the risk of suppressing the very trauma that they are trying to address. Such memorials—places where we are meant to remember—become instead places where we actively forget.

REFERENCES

"As We Always Have." Editorial. 1995. *Daily Oklahoman* (Oklahoma City). April 23, 90A.
Collyer, Stanley. 1997. "The Search for an Appropriate Symbol: The Oklahoma City Memorial Competition." *Competitions* 7 (Fall): 4–15.
Doss, Erika. 2010. *Memorial Mania: Public Feeling in America.* Chicago: University of Chicago Press.
Helphand, Kenneth. 2006. *Defiant Gardens: Making Gardens in Wartime.* San Antonio, TX: Trinity University Press.

Hunt, John Dixon. 2001. "'Come into the Garden, Maud': Garden Art as a Privileged Mode of Commemoration and Identity." In *Places of Commemoration: Search for Identity and Landscape Design*, ed. Joachim Wolschke-Bulmahn, 9–24. Washington DC: Dumbarton Oaks Research Library and Collection.

Linenthal, Edward T. 2001. *The Unfinished Bombing: Oklahoma City in American Memory*. Oxford, UK: Oxford University Press.

McMullen, Harry. 1995. "I, Too, Lost a Child in the Oklahoma City Bombing—One That I Will Never Forget." *Rocky Mountain News* (June 18).

National Park Service. 2011. *Oklahoma City National Memorial Self-Guided Tour*. Washington, DC: U.S. Department of the Interior. www.nps.gov/okci/planyourvisit/self-guided-tour.htm#CP_JUMP_658400 (accessed November 13, 2011).

O'Connell, Kim A. 2000. "The Gates of Memory." In *Landscape Architecture* (September): 71.

Oklahoma City National Memorial and Museum. 1996. *Final Report of the Murrah Federal Building Memorial Task Force, Memorial Ideas Input Subcommittee* (submitted March 1, 1996). Robert M. Johnson Collection, 249/4421 B2F1. Oklahoma City: Oklahoma City National Memorial and Museum.

———. 2011. *Outdoor Symbolic Memorial Walking Tour* (MPEG file). Oklahoma City: Oklahoma City National Memorial and Museum. www.oklahomacitynationalmemorial.org/secondary.php?section=2&catid=230 (accessed November 13, 2011).

Rossant, James, and Richard Scherr. 1997. *Oklahoma City Memorial—An International Design Competition* (competition entry board). Oklahoma City: Oklahoma City National Memorial and Museum.

Savage, Kirk. 2006. "Trauma, Healing, and the Therapeutic Monument." In *Terror, Culture, Politics: Rethinking 9/11*, ed. Daniel J. Sherman and Terry Nardin, 103–20. Bloomington: Indiana University Press.

Sturken, Marita. 2007. *Tourists of History: Memory, Kitsch and Consumerism from Oklahoma City to Ground Zero*. Durham, NC: Duke University Press.

U.S. Congress. House of Representatives, Subcommittee on National Parks and Public Lands. 1997. "Statement of Robert M. Johnson, Chairman, on behalf of the Oklahoma City National Memorial and Museum, Oklahoma City, OK." In *Hearing on H.R. 1849: To Establish the Oklahoma City National Memorial as a Unit of the National Park System . . .* , 105th Cong., 1st Session (September 9), 47. commdocs.house.gov/committees/resources/hii45289.000/hii45289_0f.htm (accessed September 15, 2011).

Weber, Hanno, Kathleen Hess, and Michael Maher. 1997. *Oklahoma City Memorial—An International Design Competition* (competition entry board). Oklahoma City: Oklahoma City National Memorial and Museum.

Young, James E. 2006. "The Stages of Memory at Ground Zero." In *Religion, Violence, Memory and Place*, ed. Oren Baruch and J. Shawn Landres, 214–34. Bloomington: Indiana University Press.

ALAN E. LONDON

Honoring Korean War Veterans

CONFLICTING VALUES OF COMMEMORATION

The Korean War Veterans Memorial (KWVM) on the National Mall in Washington, D.C., dedicated in 1995, not only commemorates a war and its military casualties but also reflects the conscious goal of its sponsors to inspire in visitors, well into the future, an emotional investment in the ongoing worth, honor, and desirability of military service. The KWVM's primary aesthetic idea—a group of free-standing figural sculptures installed without a mediating plinth or base—is unique among Washington's monuments. As a reaction to the minimalism and muteness of the Vietnam Veterans Memorial (1982), the KWVM continued the populist battle against perceived elitism. The retired generals and colonels who controlled the KWVM's development insisted that the landscape of the memorial be designed to engage visitors in ways to ensure that the values of military service would be communicated and understood. In a convoluted and contentious design process, the KWVM's designers and governmental approvers worked through multiple proposals to accomplish that engagement. While the resulting landscape is often described as successful, there are several design and programmatic difficulties that may distract visitors from the full reception of those values.

Beginnings

The Korean War was a bloody conflict fought between 1950 and 1953. Often called the "Forgotten War," it remained uncommemorated on the National Mall until the early 1980s, when Korean War veterans and their supporters were galvanized into action by the impending Vietnam Veterans Memorial (VVM) project. The political controversy surrounding Maya Lin's competition-winning design for the VVM, in which "powerful conservatives . . . lambasted Lin's design as dishonorable and worked hard to stop it from becoming realized" (Savage 2009, 276), was echoed in the early history of the KWVM.

Control of the initially private organizing committee for a Korean War memorial, incorporated in 1981 by a Korean-born naturalized U.S. citizen,[1] was soon "captured by supporters of the New Right in an organizational coup" led by Myron McKee, an unpaid staffer on the Reagan-Bush transition team (Hagopian 2012, 218).[2] At the unveiling ceremony for the Vietnam Veterans Memorial in 1982, McKee's committee handed out about six thousand questionnaires, asking veterans for their "advice and participation" in deciding what the Korean Memorial would look like. A representative of the committee noted that the questionnaire respondents were actually voicing their "disapproval of the VVM, opting instead for a traditional monument, above ground and based on what veterans want" (Conconi 1983).[3]

In 1985, faced with media criticism of the McKee committee's fundraising practices,[4] Congress determined that the Korean Memorial would be a government-controlled project. Construction of the KWVM would be carried out by the American Battle Monuments Commission (ABMC), whose primary responsibility theretofore was for American military cemeteries overseas and whose self-described purpose is to "honor the service, achievements and sacrifice of the U.S. Armed Forces" (American Battle Monuments Commission 2012). Further, Congress mandated an advisory board of Korean War veterans to decide on the design of the memorial and to be responsible for fundraising. Thus the people whose service was to be commemorated would decide what the commemoration would look like and would be responsible for getting it built.[5]

Mandated Values and Design Responses

The advisory board, whose chairman, Richard Stilwell, and vice-chairman, Raymond Davis, were retired four-star generals, held a design competition. The instructions to competitors were unambiguous: the veterans wanted a patriotic antidote to the Vietnam Veterans Memorial. They wanted a forward-looking monument that would not only honor service in the Korean War, but would also, by clear implication, honor future military service. The instructions to competitors exhorted them to design a memorial that would both express gratitude to Korean War veterans for their spirit of service and willingness to sacrifice, and remind visitors that those same patriotic values would be needed by the nation in "future emergencies." Patrick Hagopian has commented that this linkage of past and future, with its emphasis on the value of military preparedness, was "much more explicitly ideological than the program for the Vietnam Veterans Memorial . . . [which] had called for a memorial that would eschew all political statements and thus begin a 'healing process' by transcending political issues" (2012, 221).

The KWVM competition instructions permitted, but did not require, a list of names of those who died or were missing in action as an element of the design. However, any "depiction of names must engender a feeling of respect and pride rather than grief" (Korean War Veterans Memorial National Design Competition 1988), a clear reflection of the advisory board's distaste for the "victims' monument" aura of the VVM and their view that, in General Stilwell's words, "[u]nlike Vietnam, the end result [of the Korean War] was victory in geopolitical terms, with immensely favorable consequences" (Korean War Veterans Memorial Advisory Board 1989). And the inclusion of an American flag, so controversial in the VVM case, was mandated.[6] Thus the KWVM sponsors viewed the memorial as being about the Korean War in particular and military values in general. Chairman Stilwell wrote that, in the process of evaluating all the competition entries, the advisory board jury kept in mind that, "[w]hile the Korean War was the centerpiece, the Memorial provided a God-given opportunity to render an enduring salute to all Americans who have rallied to the colors, donned uniform and demonstrated the collective discipline and fortitude which has seen our nation safely through other crises" (Korean War Veterans Memorial Advisory Board 1989).

The winning proposal was submitted by a team of four design professors from the Pennsylvania State University—two architects and two landscape architects.[7] In preparing their design, the team members were inspired by David Douglas Duncan's famous photographs of lines of soldiers moving across the frozen Korean landscape and by recollections of veterans who "knew the country with their feet" (Blakely 1990, 6). The design, chosen by the advisory board solely

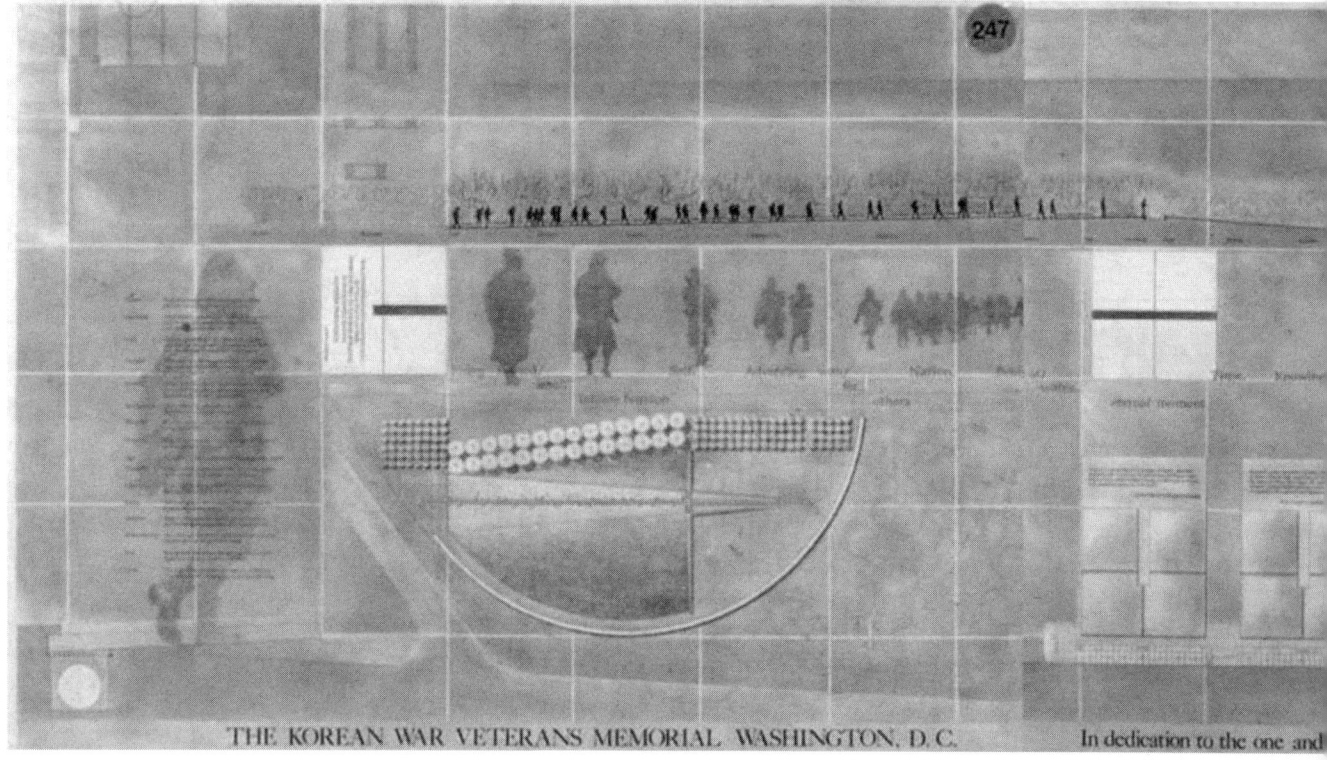

The Korean War Veterans Memorial, Washington, D. C.   In dedication to the one and

Fig. 7.1. Winning design for the Korean War Veterans Memorial. Design by John Paul Lucas, Don Alvaro Leon, Veronica Burns Lucas, and Eliza Pennypacker Oberholtzer, 1989. RG 117.3.3, National Archives at College Park, MD.

on the basis of presentation boards (fig. 7.1), featured a column of thirty-eight soldiers sculpted in granite advancing up a constructed hill toward the required American flag. The design team sensed that figurative military sculpture would be appealing to the jury of veterans (Pennypacker 2011). The soldiers would be larger than life— metaphorically the size of their courage (Johnson 1990, 71). Visitors would pass between the troopers on a narrow path, taking a ritual journey "analogous to a religious procession" (70).[8] As was made clear by the design team immediately after the competition, but was not obvious to the

jury (Lecky 2012, 55), the design was heavy with symbolism: the number thirty-eight represented the Thirty-eighth Parallel, which remains the armistice line separating North and South Korea; the plantings of barberry and tortuously pruned plane trees suggested the blood and guts of war; the arrangement of the sculpted figures functioned as a timeline, expressing victories and defeats by the postures of the sculptures and their relative proximity to one another (Burns Lucas et al. 1989). A model of the winning design (fig. 7.2) was unveiled at the White House on Flag Day, 1989, in a Rose Garden ceremony presided over by President George H. W. Bush.

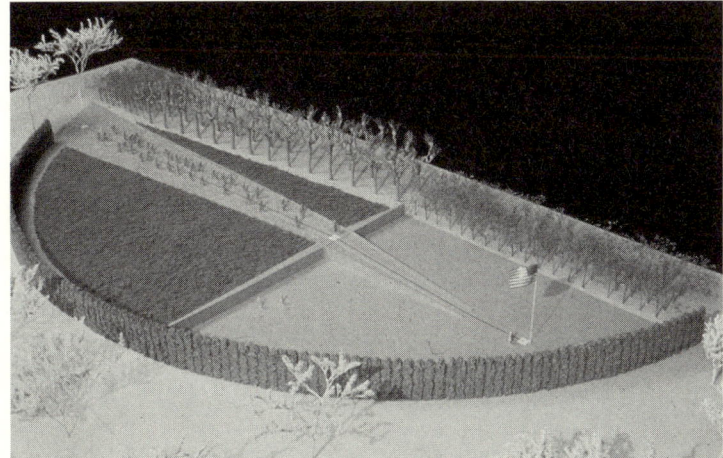

Fig. 7.2. Model of the original 1989 design for the Korean War Veterans Memorial (Lucas, Leon, Lucas, and Pennypacker Oberholtzer). From Cooper, n.d.

After reviewing the winning design, the National Park Service and the National Capital Planning Commission reacted negatively, especially to the plantings, which were seen as closing the memorial off from views of the Potomac and separating the memorial visually from the other monuments of the Mall (National Park Service 1989; National Capital Planning Commission 1989, 9). The Commission of Fine Arts (CFA), especially Chairman J. Carter Brown, disagreed. Brown remarked that Washington suffered by being a city of space, but not spaces, and that, to be effective as space-making, the design needed to be set off by

something solid. However, the CFA had other criticisms of the design (Commission of Fine Arts 1989).

The advisory board was troubled by the Penn State team's adamant refusal to change anything in response to the criticisms and by the subtlety of the symbolism and the timeline concept. The generals saw the winning design as depicting a battle formation (Johnson 1990, 70–71) and decided to replace the linear narrative device of a timeline with the more straightforward idea of a "moment in time," a narrative concept in which past and future events can be inferred from what is physically portrayed (Potteiger and Purinton 1998, 7). Concerned about the impasse with the Penn State team, the ABMC hired Cooper-Lecky Architects, an architect/engineering firm whose principals

Fig. 7.3. Korean War Veterans Memorial, aerial view.
Cooper-Lecky Architects, designers, 1995. Photo from
Carol Highsmith Collection, Library of Congress.

were W. Kent Cooper and William P. Lecky, to serve as the architects of record on the project, a role they had also played on the Vietnam Veterans Memorial. The Penn State team felt frozen out of the design process; they eventually sued the government—and lost.[9] Although a few of the elements of their winning concept remain in the as-built memorial, the Penn State team believed that the extensive changes destroyed the integrity of their design (Pennypacker 2011).

Over many months, the Cooper-Lecky team proposed refinement after refinement of their "moment in time" plan, including reducing the number of sculptures to nineteen, installing them on a triangular "Field of Service," and substituting cast stainless steel for granite. One critic has described the result of their work—the memorial we see today (figs. 7.3 and 7.4) with its round black pool and surrounding linden bosque pierced asymmetrically by the triangle—as an "arresting compositional tension [of] the softly molded soldiers [against] a highly abstracted, almost Platonic landscape" (Lewis 1996, 17).[10]

Fig. 7.4. View of the memorial from the west. Cooper-Lecky Architects, designers; Frank Gaylord, sculptor, 1995. Photo from Carol Highsmith Collection, Library of Congress.

Arguments over what the sculpted troopers should look like were among the most contentious of the design approval process and could be viewed as evidence of a populist/elite binary with respect to values in commemorative spaces. The advisory board wanted the troopers' uniforms to be rendered in "exquisite detail," instructing that "all weaponry, gear, and uniforms should be historically exact" so that visitors could understand what it meant for a person to be a combatant in the Korean War (Korean War Veterans Memorial Advisory Board 1990). In response, CFA Commissioner Robert Peck expressed reservations about the wealth of informational material proposed, raising an issue about the interplay of educational and emotional values that, in his view, transcended the KWVM. Referring to the FDR Memorial and to the Persian Gulf War, which had just begun, Peck worried about "setting a precedent. . . . Our memorials seem to be turning into outdoor museums" (Gamarekian 1991). Discussing the details of race, rank, service, and function to be reflected in the sculptures, Chairman Brown suggested that a "Disney World approach where one is being instructed and one is into info-tainment" is inappropriate to the "authenticity" of the environment of Washington's monumental core (Forgey 1991).

Although disagreement about the value of education in commemoration seems like the primary issue here, another analysis may be possible. For the advisory board, focusing on the details of insignias, uniforms, and equipment was a way of reifying the "spirit of service" they were intent on commemorating. Stilwell spoke of saluting those who "donned uniform." The uniform is a powerful symbol, and the distinctions among uniforms can elicit and enhance feelings of identification and pride. One need only walk the streets of a city like Pittsburgh during football or hockey season to understand how a uniform can generate and express emotion. For some people, tangible symbols, particularly those that provide the intimacy of being worn on the body, are necessary or helpful in achieving emotional involvement. For others, such symbols are unnecessary or even counterproductive. It may be too much to interpret these alternatives as "highbrow" and "lowbrow," but certainly there are different modes of receiving and processing environmental cues that were at play in the process of designing the KWVM.[11]

Visitor Engagement and Distractions

In *Narrative Landscapes: Design Practices for Telling Stories*, Potteiger and Purinton describe "how space and form can dramatize and exert control over the sequential unfolding of a story" (1998, 12). Critics have argued that the as-built KWVM fails because (unlike the competition-winning design) it provides no opportunity for ritual, no "processional experiences through space" (Woland 2005, 45) of the kind experienced at the Lincoln and Vietnam Veterans memorials and especially at the FDR Memorial, where the commemorative design concept of a ritual procession was carefully studied, applied, and described by its designer, the landscape architect Lawrence Halprin (1997, 20). Certainly, it can be argued that the designers have created a "place apart" at the KWVM and that the path visitors take from behind the troopers and past (or metaphorically among) them to the apex of the Field of Service creates an element of drama. However, the flat, compact, and overlapping quality of the ground plane deadens any sense that the visitor's movement is purposeful: there are no steps to climb or corners to turn, no hint of surprise or expectation of an unseen destination. The memorial's "compositional tension" creates a meaningful experience, but it is an experience of looking, of relating, of appreciating, of thinking, and of honoring, rather than an experience of a ritual procession through space.

The fact that people enter and exit the memorial at either of two points, at the northwest and southwest corners of the triangle's base, contributes to the visual degradation of a sense of ceremony. Most tour buses park closer to the southwest entrance, which is also closer to visitor parking on Ohio Drive. As a result, many visitors miss the experience of first seeing, at the northwest entrance, Trooper No. 17 (figs. 7.5 and 7.6), with his evocation of danger, or perhaps of sacred space, that is a major narrative element of the

Fig. 7.5. Trooper no. 17, facing northwest entrance to the memorial. Photo by the author, 2011.

memorial. Further, many visitors leave on the same path on which they entered, missing the experience of the mural wall from either close-up or farther away. Most important, the idea of a ritual journey is negated by the people always walking past you in the opposite direction. Somehow, at the Lincoln and Vietnam memorials, encountering people who are descending as you are climbing (or vice-versa) reinforces the processional feeling: there is a bond with strangers who have just been where you will soon go or who are about to go where you've just been. At the KWVM, on the other hand, there is no ascent or descent; the slope seems even flatter than it is, so the two-way traffic on the paths seems more like a city sidewalk. Despite their myriad other problems, earlier versions of the design, both the competition winner and Cooper-Lecky's first revision, would potentially have solved this problem by imposing a one-way flow on a steeper slope.[12]

Although the circulation pattern at the Korean War Veterans Memorial has been praised for its "clarity," its very compactness allows the visitor's commemorative experience (including an openness to value messages) to be marred by distractions that are not part of the design. Two of these distractions—which should, perhaps, have been predicted and mitigated in the design process—are photography and floral tributes.

The memorial is highly photogenic. Each one of the nineteen sculpted faces holds perfectly still for as long as it takes the pickiest photographer to get multiple perfect portraits. The postures and interactions of the troopers, the reflections on the mural wall and in the circular pool, the multiple viewpoints for each design element—all encourage photographers by providing appealing challenges to the amateur's image-composing ability.[13] On the National Mall, only the FDR Memorial competes with, and perhaps trumps, the KWVM in its multiplicity of opportunities for photographic creativity. This isn't surprising, for both memorials are garden landscapes in which multiple naturalistic elements are employed to create complex narratives, and both are sandwiched, on the timeline of National Mall commemorations, between the minimalism of the VVM and the cold monumentality of the World War II and Martin Luther King memorials. Because of the diffusion of symbolic episodes in Halprin's FDR Memorial, there is plenty of room for visitors to compose shots without the constant threat of other visitors moving through the frame. The opposite occurs in the KWVM's confined space, where every step a visitor takes can ruin another visitor's photo opportunity. The pedestrian traffic jams at the apex are reminiscent of a six-way vehicle intersection where no one knows who has the right of way. The comedic "after you, Alphonse,"

OPPOSITE:
Fig. 7.6. Detail of Trooper no. 17. Photo from Carol Highsmith Collection, Library of Congress.

routines among photographers can't be the kind of value-centered visitor interaction the designers and sponsors were hoping for.

The sense of experience shared with people one doesn't know is an essential element in the patriotic values the sponsors wanted the KWVM to impart or elicit. As Kirk Savage (2009, 275) has described in his discussion of the "participatory" quality of the VVM experience, one receives that sense walking along the Vietnam wall, notwithstanding the absence of intentional value statements. But, ironically, it can be missing in the experience of the KWVM. Picture-taking, like talking on a cell phone, excludes the strangers around you (from your concentrated activity, if not from your photo), and the surfeit of photography at the KWVM exacerbates the feeling of disconnectedness with fellow visitors.

Writing similarly about Thiepval, the Edwin Lutyens–designed memorial in northern France to British soldiers missing in World War I, Savage notes that, while most commentators focus on the visitor's personal elegiac experience, that monument should also be seen as "a shared space, a collective experience shared both synchronically (with others at that very moment) and diachronically (with all those who have come before, deposited wreaths and flowers long since withered)" (2010, 649).

Many floral wreaths have been "deposited" at the Korean War Veterans Memorial over the years, and the flowers are seldom allowed to wither. Expanding the longstanding tradition of placing commemorative wreaths on war memorials at various occasions during the year, the Republic of Korea Embassy and other South Korean groups arrange for continuous elaborate floral displays to communicate the continuing Korean gratitude for U.S. war efforts. The colorful wreaths are placed at the apex of the Field of Service, in front of the Lead Trooper (fig. 7.7). While this gratitude echoes the mandate of the authorizing statute, the wreaths may nevertheless be seen as programming that is detrimental to the memorial's function as a communicator of values and distracting from its aura, from the sense of coming face-to-face with representations of troopers who are in danger.

First, the floral tributes are red-white-and-blue intrusions on the muted gray-and-black color scheme of the memorial's hardscape, invoking the "black and white war" that folks back home saw on early 1950s television and newsreels (Lecky 2011). Further, they disrupt the sense of ghostliness of the sculptures and haziness (as in the "memory's haze" invoked in Yale's alma mater) of the mural photographs. Second, placement of the wreaths blocks the visitor's visual interaction

OUR NATION HONORS
HER SONS AND DAUGHTERS
WHO ANSWERED THE CALL
TO DEFEND A COUNTRY
THEY NEVER KNEW
PEOPLE

Fig. 7.7. Floral tributes. Photo by the author, 2010.

with the troopers, which the sponsors and designers considered essential to full emotional and intellectual engagement.[14] Third, is it possible that a diachronic collective experience of the kind Savage describes could be diminished by the wreaths bearing inscriptions that non-Korean-speaking visitors can't understand? Might the visitor be confused by the juxtaposition of the wreaths with the inscription, just below the wreath stands, praising American soldiers who defended a "country they never knew and a people they never met"?

Conflicts Renewed

In 2013, and again in 2015, a congressional bill has proposed to construct at the memorial a transparent glass "wall of remembrance"

The Wall of Remembrance

A proposal for a glass wall surrounding the plaza around the Pool of Remembrance at the Korean War Veterans Memorial. Etched in the transparent glass panels, and softly glowing in the twilight, are the names of the fallen and expressions of gratitude and sacrifice by the people of South Korea.

Design/Concept : Architect WILLIAM P. LECKY - FAIA
Digital Imagery : Harold H. Gonzalez

National Mall,
Washington, DC. April, 2008

KOREAN WAR MEMORIAL · WALL OF REMEMBRANCE

Fig. 7.8. Rendering of proposed "wall of remembrance" at the memorial. William P. Lecky, designer, and Harold H. Gonzalez, digital designer, 2008. Courtesy of William P. Lecky, FAIA.

on which would be inscribed the names of all American military personnel who died in the Korean War. The veterans and others who support the bill and the proposed design (fig. 7.8) argue that the KWVM has failed in one of its congressionally mandated purposes: to express adequately the nation's gratitude to those who paid the price of freedom in Korea (U.S. Congress, Weber 2011). The National Park Service opposes the bill, arguing that the memorial is a completed work of public

art[15] and that the glass wall would violate the Commemorative Works Act (U.S. Congress, Whitesell 2011).

Although one of the surviving members of the advisory board is the leader of the effort to build the wall of remembrance (U.S. Congress, Weber 2011), the project seems inconsistent with Stilwell's primary goal, the ongoing expression of the value of military service, with its corresponding commemorative program which uses the rhetorical device of synecdoche[16] to create a narrative of the "universal" soldier. Unlike the Vietnam Memorial, the whole thrust of the KWVM's as-built design is commemoration through recognition, not commemoration through elegy. The memorial's narrative honors veterans, including the dead, as actors, not as victims. The glass wall, when added to the troopers and the mural wall and the pool and the didactic inscriptions, could undercut both the memorial's essential narrative story—which is not that "people died" in Korea but that "people served" in Korea—and the memorial's essential value-laden message that future generations should honor military service to the nation and be inspired to replicate it in the future.

Equally important to landscape design analysis is assessing the proposed wall of remembrance in the context of the National Mall. It has been said that Washington's monuments immerse visitors in the "essential" America, the "soul of the nation" (Reston 1995), and that the memorial landscape of the Mall creates an "abiding sense of national identity" (Savage 2009, 10). Each of the three major war memorials on the western half of the Mall accomplishes this evocation in its own way, with its own narrative scheme. The World War II Memorial is a memorial of places—states, territories, the Atlantic and Pacific theaters, the Hürtgen Forest, Bataan. The Vietnam Veterans Memorial is, of course, a memorial of names. And the Korean War Veterans Memorial is a memorial of faces, sculpted and etched.

Whether intentional or not, the distinction among these three modes of commemoration adds an element of complexity and richness to the Mall visitor's experience that isn't provided by the monuments to the great men, no matter how effective those may be otherwise. Each of the Lincoln, Jefferson, Roosevelt, and King memorials features a monumental figurative sculpture of the subject's likeness and inscriptions of his words. After Lincoln's, which was the first of these, any other typology could have seemed somehow inappropriate or even disrespectful. On the other hand, the disparate design treatment of the three war memorials reflects, perhaps, the ambiguities and contradictions inherent in Americans' attitudes about war. Even citizens who hate war can comfortably honor those who fought and maybe died in one, but there is value in preserving different modes of expressing and inviting that commemoration. In this sense there is now a narrative equipoise—a dialectic or tension—among the three war memorials that circle the great space of the Reflecting Pool. The proposed glass wall of remembrance at the Korean Memorial would disrupt that balance.[17]

The author is grateful to Barry Schwartz for graciously making available an unpublished essay, to Eliza Pennypacker and Bill Lecky for cheerfully and expansively submitting to lengthy interviews, to Patrick Hagopian for generously sharing research materials and ideas, and to Kyle Beidler for serving inspirationally as thesis advisor and mentor.

1. Dr. Chayon Kim, the daughter of a wealthy Korean landowner and former curator of a U.S. military museum in Seoul, incorporated a "national committee" to finance and construct the first permanent memorial in Washington, DC, to Americans who served in the Korean War (Bowers 1985).

2. Kim hired McKee as executive director of her committee. McKee's "primary job was to convince Congress" to set aside federal land in Washington as the site for the memorial (Bowers 1985) and, by June 1982, a bill to approve a Korean War Memorial was introduced in both the House of Representatives and the Senate (H.J.Res. 523, 97th Cong., 2nd Sess., 1982). In her appearance before the U.S. Commission of Fine Arts in November 1982, Dr. Kim stressed the role of the United Nations in the Korean conflict and commented: "the Korean War has been the only one thus far to be fought under the flag of the United Nations" (Commission of Fine Arts 1982). A revised bill, adding a reference to honoring "the allied forces" as well as U.S. forces, was introduced in Congress in April 1983 (H.J.Res. 236, 98th Cong. 1st Sess., 1983). However, there is nothing similar in the statute as eventually enacted (Pub.L. No. 99-572, 100 Stat. 3226, 1986) about honoring the forces of other UN countries.

3. Although McKee's committee had no government mandate to sponsor a Korean War memorial, it published a prospectus stating that competition entrants could be companies or individuals, including professionals and students, but that "draft evaders or those with discharges under other than honorable conditions need not apply" (Furgurson 1983b).

4. The Washington bureau chief of the *Baltimore Sun*, Ernest Furgurson, wrote a series of articles exposing what he said were illegal and unethical practices by the McKee group (Furgurson 1983a; 1983b; 1984). A House of Representatives committee held hearings, and the IRS, the FBI, and the U.S. Postal Service all conducted investigations of McKee's committee, but no charges were ever filed (Bowers 1985). Nevertheless, many contributors and volunteers felt victimized by the committee (Barker 2011).

5. By contrast, Maya Lin's design for the VVM had been selected by a jury of art and design professionals (including the editor of *Landscape Architecture* magazine), none of whom was a Vietnam veteran (Hagopian 2009, 461–62).

6. As Barry Schwartz has pointed out (n.d., 9), members of the advisory board were offended by the Vietnam Veterans Memorial's muteness and lack of a patriotic message. Because the Commission of Fine Arts (CFA) had approved the composition of the VVM jury, the CFA was viewed by critics of the VVM as responsible for what they didn't like about it. Schwartz concludes that, by establishing the advisory board as the jury for the first national commemoration of the Korean War, Congress decided that veterans should have the opportunity to "define" the war and, by establishing the CFA as one of the governmental agencies whose approval would be required before the memorial could be built, the act made it likely that there would be a battle over that definition.

7. The architects were John Paul Lucas and Don Alvaro Leon; the landscape architects were Veronica Burns Lucas and Eliza Pennypacker Oberholtzer. After Ms. Oberholtzer, in the third trimester of pregnancy (Pennypacker 2011), withdrew from active participation in the project, the remaining three designers styled themselves "BL3" and undertook design development, then the subsequent lawsuit.

8. The context of Johnson's analogy is the designers' own design statement, presented at their first meeting with the advisory board after the competition judging, which "refers diagrammatically to the plan of Reims Cathedral as an 'invisible geometric construct.' . . . The visitor enters through the narthex (the bosque of dogwood tress), advances among the soldiers, through a narrow, forced perspective up the nave to the crossing, where the north transept points toward the Vietnam Veterans Memorial. . . . The passage through the soldiers is neither accommodating nor comfortable, but why should war memorials be easy to visit?" (Johnson 1990, 70–71).

9. It is tempting to characterize the conflict between the advisory board and BL3 as a continuation of the populist/elitist battles surrounding the VVM. However, the impression one gets from the record is that BL3 was on board with the value messages the advisory board wanted the memorial to convey. Indeed, Benjamin Forgey, the architecture critic of the *Washington Post,* wrote in an initial review of the design that it "tells a story of soldiers motivated by patriotism. . . . It is a story such as generals would like to hear again and again . . . a nostalgic and uni-dimensional celebration of a citizen army obeying duty's call (and orders), but excluding other interpretations" (Forgey 1989). Rather, as William Lecky, of Cooper-Lecky, has noted, "there was a kind of instant friction that began to occur between BL3 and the Advisory Board" (Lecky 2011). Like Maya Lin, BL3 argued against any changes to their competition-winning design. But while the simplicity of Lin's design made stark the choice to support or oppose it, the complexity of BL3's effort gave everyone—elite and populist alike—something to complain about.

10. Intent on honoring all U.S. military functions, the advisory board insisted that support troops be well represented in the narrative, which the designers accomplished by laser-etching actual photographs of anonymous thousands of such troops onto a granite wall flanking the soldier sculptures. The record of these design developments reflects much specific criticism, and specific design advice, from J. Carter Brown (Kohler n.d., 208; Lecky 2012, 83), notwith-standing the CFA chairman's frequent protestations that he couldn't design the memorial.

11. The designers solved this problem brilliantly by deciding that the troopers should be sculpted wearing wind-blown ponchos. This not only allowed the controversial uniform details to be mostly covered up, it also helped the narrative by suggesting the foul weather of Korea and helped unify aesthetically the composition of nineteen separate figures (Lecky 2012, 76).

12. The "moment in time" scheme also has a role in undermining a sense of procession and ritual. In the memorial's narrative, the moments before and after are implied and left to the visitor's educated imagination. It is hard to conceive of a place as a metaphor for a journey when the place portrays, figuratively, a single moment, even if the moment lasts as long as a visit to the place.

13. In this respect, the KWVM is almost unique among the memorials on the Mall. The Washington Monument looks pretty much the same in every photo. There's one best angle from which to photograph the Lincoln Memorial, and visitors, dwarfed from that angle, can ameliorate the composition. For the Jefferson Memorial, it can be argued that there are two best angles—one straight through the portico with Mr. Jefferson silhouetted against the sky, the other from across the Tidal Basin, framed by cherry blossoms. Much of the photographic effort at the Martin Luther King Memorial seems aimed at getting the kids' faces and Dr. King's face recognizable in the same frame. One observes many fewer photographs being taken at the VVM than at the other memorials: perhaps the quiet, reverent mood there makes picture-taking seem less appropriate.

14. The intensity of the sculptural group's faces and bodies, at the very point where one first gets to see

the platoon face on, is partially obscured by a frothy assortment of dyed carnations, mini-flags, pinwheels, and ribbons. Installed directly on the planted base plane, emerging from the zelkova woods without a mediating plinth or base to frame the representational scene, the sculptural group is particularly vulnerable to the floral distraction.

15. A similar argument was made against Frederick Hart's figural sculpture of three servicemen at the VVM. The argument was ultimately unsuccessful, but the sculpture was placed at some distance from the Wall as a compromise (Savage 2009, 277).

16. Potteiger and Purinton define synecdoche as "the use of a part of something to represent the whole, or of the whole to stand for the part," noting that it is a "particularly effective device in landscape narrative because it can conjure up a whole complex story just by using a piece or fragment from the story" (1998, 37).

17. And yet the urge to add names to faces—and faces to names—is powerful. The current plan for the VVM's proposed underground "education center" includes a display of oversized portrait photos, to be supplied by loved ones, of the war dead whose names are inscribed on the wall. Each photo would be displayed on only one day a year—the subject's birthday. It's hard to imagine any concept more diametrically and disconcertingly opposed to Maya Lin's VVM design competition entry statement that "death is in the end a personal and private matter, and the area contained within this memorial is a quiet place meant for personal reflection and private reckoning" (Lin 2000, section 4, p. 5).

REFERENCES

American Battle Monuments Commission. 1989. "545 Designs Submitted to the American Battle Monuments Commission for the Korean War Veterans Memorial" (color slides), 1988–89. Record Group 117.3.3. National Archives at College Park, College Park, MD.

Barker, Hal. 2011. "Korean War Veterans Memorial Story." *Korean War Project*. www.koreanwar.org/html/korean_war_veterans_memorial_s.html (accessed July 20, 2011).

Blakely, Stephen. 1990. "Remembering the Forgotten War." *Research/Penn State* (June).

Bowers, Rick. 1985. "Fight Within Hurts Effort for Korean War Memorial." *Miami Herald* (July 7).

Burns Lucas, V., D. A. Leon, J. P. Lucas, and E. Pennypacker Oberholtzer. 1989. "Design Statement of the Korean War Veterans Memorial." Transcript of the meeting of the National Capital Planning Commission (July 27, 1989). Record Group 328.2. National Archives Building, Washington, DC.

Commission of Fine Arts. 1982. "Minutes of a Meeting of the Commission" (November 10). Files of the Commission, Washington, DC.

———. 1989. "Letter from Commission Chairman J. Carter Brown to Robert Stanton, Regional Director, National Park Service" (August 9). Files of the Commission, Washington, DC.

———. 1991. "Minutes of a Meeting of the Commission" (February 21). Files of the Commission, Washington, DC.

Conconi, Chuck. 1983. "Personalities." *Washington Post* (June 9).

Cooper, W. Kent. N.d. "The Role of Design Competitions in Shaping the West End of the National Mall," illustrations accompanying presentation (accessed ca. July 24, 2011; website now discontinued).

Duncan, David Douglas. 1951. *This Is War!* New York: Harper & Brothers.

Forgey, Benjamin. 1989. "How Many More Memorials? Korean Vets, Police Officers Designs Lead a Tight Commemorative Field." *Washington Post* (June 17).

———. 1991. "War Memorial Design Rejected; Commission Wants Korean Monument Scaled Down." *Washington Post* (January 18).

Furgurson, Ernest B. 1983a. "Another War Monument Is Needed 'To Rectify an Old Injustice.'" *Baltimore Sun* (January 7).

———. 1983b. "Financing a Korean Memorial." *Baltimore Sun* (August 16).

———. 1984. "Korean War Memorial: Questions Continue, Congressional Probe Is Needed." *Baltimore Sun* (September 2).

Gamarekian, Barbara. 1991. "Panel Turns Down Plan for Korean War Shrine." *New York Times* (January 18).

Hagopian, Patrick. 2009. *The Vietnam War in American Memory: Veterans, Memorials and the Politics of Healing.* Amherst: University of Massachusetts Press.

———. 2012. "The Korean War Veterans Memorial and Problems of Representation." *Public Art Dialogue* 2 (2): 215–53.

Halprin, Lawrence. 1997. *The Franklin Delano Roosevelt Memorial.* San Francisco: Chronicle Books.

Johnson, Jory. 1990. "Granite Platoon." *Landscape Architecture* 80 (1): 69–71.

Kohler, Sue A. N.d. *The Commission of Fine Arts: A Brief History, 1910–1995.* Washington, DC: Commission of Fine Arts.

Korean War Veterans Memorial Advisory Board. 1989. "Minutes of the July 21 Meeting of the Advisory Board Executive Committee." Record Group 220.19.15. National Archives at College Park, MD.

———. 1990. "Enclosure #1—Board Position on the 38 Statues" (January 31). Record Group 220.19.15. National Archives at College Park, MD.

Korean War Veterans Memorial National Design Competition. 1988. "Design Competition Description and Rules." Files of the Commission of Fine Arts, Washington, DC.

Lecky, William P. 2011. Personal interview by Alan London (May 17). Cooper Lecky Architects, McLean, VA.

———. 2012. *Designing for Remembrance: An Architectural Memoir.* McLean, VA: LDS Publishing.

Lewis, Roger K. 1996. "Washington Monuments: 'Battles over the Mall.'" *Architectural Record* (January): 17–21.

Lin, Maya. 2000. *Boundaries.* New York: Simon & Schuster.

Lowenthal, David. 2002. "The Past Is a Theme Park." In *Theme Park Landscapes: Antecedents and Variations,* ed. Terence Young and Robert Riley, 11–23. Washington, DC: Dumbarton Oaks Research Library and Collection.

National Capital Planning Commission. 1989. "Draft Report to the National Park Service and the American Battle Monuments Commission" (July 20). Included in the transcript of the meeting of the National Capital Planning Commission, July 27, 1989. Record Group 328.2. National Archives Building, Washington, DC.

National Park Service. 1989. "Letter of Ronald N. Wages, Acting Regional Director, National Capital Region, to J. Carter Brown" (July 17). Files of the U.S. Commission of Fine Arts, Washington, DC.

Pennypacker, Eliza. 2011. Personal interview by Alan London (February 24). Pennsylvania State University, State College.

Potteiger, Matthew, and Jamie Purinton. 1998. *Narrative Landscapes: Design Practices for Telling Stories.* New York: John Wiley & Sons, Inc.

Reston, James Jr. 1995. "The Monument Glut." *New York Times* (September 10).

Savage, Kirk. 2009. *Monument Wars—Washington, DC, the National Mall, and the Transformation of the Memorial Landscape.* Berkeley: University of California Press.

———. 2010. "The War Memorial as Elegy." In *The Oxford Handbook of Elegy,* ed. Karen Weisman, 637–657. New York: Oxford University Press.

Schwartz, Barry. N.d. "The Revenge of Tradition: Erecting the Korean War Veterans Memorial." Unpublished manuscript provided to author by

Dr. Schwartz, professor emeritus of sociology, University of Georgia.

U.S. Congress. House of Representatives, Subcommittee on National Parks and Public Lands. 2011. "Panel 2: Testimony of Col. William E. Weber" and "Testimony of Stephen E. Whitesell." *Legislative Hearing on H.R. 2563, H.R. 1335 and H.R. 845.* 112th Congress, 1st Session (October 4).

naturalresources.house.gov/Calendar/EventSingle .aspx?EventID=261983 (accessed September 14, 2012).

Woland, Jake. 2005. "The Korean War Veterans Memorial: A Confusing Memorial to a Perplexing Historic Event." *Critiques of Built Works of Landscape Architecture* 9: 43–48.

STEPHEN SEARS

Forging Commonplace

THE OLD NORTHWEST TERRITORY'S EMBLEMATIC TRANSITION FROM WILDERNESS TO LANDSCAPE

The land area once known as the Old Northwest Territory was ceded as a whole by Great Britain to the newly established United States in the Treaty of Paris (1783), the pact that ended the American War of Independence. In July 1787, the Congress of the Confederation passed the Northwest Ordinance, officially titled "An Ordinance for the government of the territory of the United States North West of the river Ohio," establishing federal sovereignty for new states and a means of governing the 260,000-square-mile territory (Havighurst 1946). It took half a century for the territory to be finally organized as the states of Ohio (1803), Indiana (1816), Illinois (1818), Michigan (1837), Wisconsin (1848), and Minnesota (1858) (Rohrbough 2008, 14), and those geographic delineations ultimately framed the region's enduring cultural and political identity.

The seemingly fixed present-day configuration of the region belies the complexity of its incremental development. Works by Walter Havighurst (1946), Henry Nash Smith (1950), and Malcolm J. Rohrbough (2008) detail the intricate timeline of the era. Forces such as competing claims between states, influences of indigenous people, and boundaries that shifted with the evidently incongruent realities of geography and politics all led to short-lived treaty lines, legislative boundaries, and proper place names that have been largely forgotten today (Sommers 1977). What did remain regionally constant during the nineteenth century were the land's seemingly endless potential and the citizenry's surety in realizing that potential by engaging in endeavors that transformed wilderness into landscape.

To apprehend their task of replacing the region's wilderness with a "productive" utilitarian landscape, inhabitants only had to be familiar with the Great Seal of the territory. The symbol conceived for the seal was inscribed with the Latin motto, "Meliorem Lapsa Locavit," which loosely translates as "a better one has replaced it," and shows a newly planted tree growing over a fallen log (Hallock 2003, 10). For the Northwest Territory this meant that the civilizing influences of future husbandmen would supplant wilderness to the betterment of the country. The territory's first governor, Arthur St. Clair, first introduced the seal by proclaiming that the elements of its iconography "all combine forcibly to express the idea that a wild and savage condition is to be superseded by a higher and better civilization" (Galbreath 1925, 553). That the phrase for the seal was borrowed from South Carolina, originally used to commemorate a victory over British forces, suggests that

influential U.S. statesmen with preconceived ambitions toward the Old Northwest Territory endeavored to use the seal as a means of indoctrinating the region's future agrarians with noble goals. That settlers were readily willing to make manifest this notion in the new territory gives credence to a deliberate and developing narrative.

The rapid domestication of the "Middle West" and its importance as a region to the early United States can be attributed to the unique circumstances of geography, natural resources, technology, and human ambition. The patterns of the region's pastoral environment are the result of both ecological systems and the purposeful practices of settlement—or nature plus culture. The push to improve the newly acquired wild lands and the later recounting of those efforts in an era when the young nation was forging its own heritage established the Old Northwest Territory agrarian landscape as a nationally significant cultural phenomenon and a key signifier in the development of a larger, particularly American mythos.

Toiling against Wilderness: Origins of an American Myth

The region's potential as an agricultural engine was set forth long before settlers arrived. A number of geologic circumstances would prove fortuitous to the prosperity of future farmers. The inland seas that formed more than 300 million years ago were responsible for the sedimentary layers and a neutral pH, chemically well-suited to crops. As recently as 10,000 years ago, advancing and receding glaciers tilled the substrate of silt, sand, and clay into an optimal growing medium. And yet, as Roderick Nash writes, "for most of their history, Americans regarded wilderness as a moral and physical wasteland fit only for conquest and fructification in the name of progress, civilization, and Christianity" (1967, n.p.). This sentiment begins to explain the midwestern landscape. The term "landscape" denotes shaped land— purposely built space—and the land formed from the Old Northwest Territory eventually would become an uninterrupted built environment from one state line to another.

In *De Natura Deorum* (45 BCE), Cicero observed, "We have also taken possession of all the fruits of the earth. Ours to enjoy are the mountains and the plains. Ours are the rivers and the lakes. We sow corn, we plant trees, we fertilize the soil by irrigation, we dam the rivers and direct them where we want. In short, by means of our hands we try to create as it were a second nature within the natural world" (1972, 184–85). In Cicero's model, first nature—wilderness—is the realm of the gods, while second nature describes the human bounds of shaped land. As David Nye puts it, "nineteenth century Americans saw no irreconcilable contradiction between nature and industry; rather, they enjoyed contemplating the dramatic contrasts created by rapid progress" (1994, 39). For example, settlers' perception of trees provides us with an insight at odds with a contemporary sense of the inherent goodness of trees. John Stilgoe notes, "Americans found those trees beautiful that indicated the most fertile soil" (1982, 145). Alexis De Tocqueville found a similar utilitarian sense among settlers when he ventured into the wilderness of Michigan in 1831. In his *Democracy in America*, written in 1835, he recalled: "Europe is much concerned with the American wilderness, but Americans themselves hardly give it a thought. The wonders of inanimate nature leave them cold, and it is hardly an exaggeration to say that they do not see the admirable forests that surround them until the trees fall to their axes. Another spectacle fills their eyes. The American people see themselves tramping through wilds, draining swamps, diverting rivers, populating soli- tudes, and taming nature. Americans do not reserve this magnificent image of themselves for rare occasions. It accompanies the most trivial as well as the most important actions of each and every one of them and is always present in the mind's eye" (1835–40, 557).

Conceptualizations of such a duality

Fig. 8.1. *George Washington in the Wilderness.* From Scott 1884, 126.

between man and environment were reinforced in portrayals of the country's founders, such as illustrations showing future presidents engaged in husbandry or fighting the forces of nature. One of the country's most enduring visages, George Washington, is shown in one image toiling as a farmer in domesticated landscapes (Anonymous 1855) and, in figure 8.1, on guard as an explorer in the treacherous wilderness (Scott 1884). In an era when the nation's narrative was still largely unwritten, dissemination of

such imagery served to focus the developing milieu. The construct of Cicero serves us with a neoclassical model to delimit the actions of early mid-American pioneers, who were sharply focused on domesticating an otherwise harsh environment. Thomas Jefferson saw such efforts as critical to the success of the newly democratic United States. In a 1787 letter to James Madison soon after the U.S. Constitution was drafted, Jefferson explains, "our government will remain virtuous for many centuries; as long as they are chiefly agricultural; and this will be as long as there shall be vacant lands in any part of America. When they get piled up upon one another in large cities, as in Europe, they will become corrupt as in Europe" (Edwards 1943, 7).

By the 1840s, when the settlement and development of the Old Northwest Territory was well underway, advocates of expansionist policies eventually framed under the expression "Manifest Destiny," such as journalist John L. Sullivan, wrote editorials supporting U.S. expansion into other western territories. Historian Henry E. Weeks has noted three key themes to such positions: (1) The virtue of the American people and their institutions; (2) the mission to spread these institutions, thereby redeeming and remaking the world in the image of the United States; and (3) the destiny under God to do this work (1996, 61). Although Manifest Destiny was part of Andrew Jackson's rhetoric of romantic nationalism, it was never official government policy. However, its tenets were well suited to the grand ambitions of pioneers "ordained" to transform the wilderness.

The lasting impression of the yeoman farmers' contribution to the development of the United States has been generally attributed to Frederick Jackson Turner's paper, "The Significance of the Frontier in American History," delivered to the American Historical Association at Chicago's World's Columbian Exhibition in 1893. Henry Nash Smith explains in his *Virgin Land* that Turner's paper influenced a whole generation of historians who rewrote American history in those terms. Smith reflects that Turner's polemic challenging accepted notions of the making of the frontier "has been worked into the very fabric of our conception of our history" (1950, 250). As a consequence, those that settled here are commemorated as yeoman farmers—free landowners, with the fiercest kind of ambition—seeking bounty and dominion against the frontier. Exemplary of such a characterization is Lorado Taft's sculpture "The Pioneers," unveiled in Taft's hometown, Elmwood, Illinois, in 1928 (Garvey 1988). The guise of the work was almost certainly influenced, even indirectly, by Turner's widely accepted ideas presented thirty-five years before in the same city where Taft maintained his studio. The portrayal of a stylized history—the heroic farmer standing *against* wilderness—would become an irresistible narrative for later cultural contributors.

Idealistic Transition to an Emblematic Landscape

When permanent settlers arrived in the Old Northwest Territory in the 1780s, it was by no means an unbroken wilderness. Long before the acquisition of the territory by the United States, there were French forts, extensive trading networks, and British settlements (Clayton 1996). Some argue that Illinois's prairie peninsula was due to Native Americans' hunting practices (Watts 1975). Indeed, prior to European intrusions, indigenous cultures had well-developed settlements that subsequently influenced the configuration of roads and the location of towns (Scharf 1900). Nevertheless, early portrayals of the territory's environment from the first half of the nineteenth century reflected an almost threatening world—one where challenges to the ambitions of future settlement were pronounced.

What is now the part-natural, part-invented boundary between Indiana and Illinois once marked the ecotone between the Eastern Hardwood Forest and the prairies of the Great Plains. As the territory gained its present-day state-line demarcations, that ecotone was erased almost completely. In the hardwood forests the task was clearing land. In Indiana alone, settlers cut an estimated 30 billion board feet of hardwood timber in the thirty-four years between 1869 and 1903 (DenUyl 1954). In Illinois and westward, it was the prairie with which the settlers had to contend. After John Deere perfected his one-piece steel plow in 1837, the settlers began to cultivate the prairie with a vengeance.

Walter Havighurst quotes a traveler reporting back to the East about his 1837 journey to Springfield, Illinois: "Our far west is improving, astonishingly. It is five years since I visited it, and the changes within that period are like the works of enchantment" (1956, 219). During the first few decades of the nineteenth century, accounts of successful ventures fed the next speculative wave into unpeopled and unimproved tracts, intensifying the already unprecedented transformation of the territory's wilderness into shaped land.

The Great Seal of the territory was only the precursor of formalized intentions toward the new expanse of natural resources now available to the eastern United States. The impulses of Manifest Destiny were indelibly codified in the official seals of states at the forefront of the country's burgeoning continental migration. In numerous examples, the iconography delineated in state seals includes the implements for subduing both wilderness and native people. The figures also explicitly illustrated the means of purposeful practice and the degree to which abundance could be reaped.

The near lack of extreme topography and the seemingly endless expanse offered a fertile ground for government-imposed order. The public land survey system, first implemented in the Seven Ranges of southeast

Ohio, would be perfected in the Old Northwest before advancing with more surety beyond the Mississippi River (Hart 1968). Jefferson suggested the system for the country after the War of Independence so that land could be parceled out to war veterans. He also considered his survey system as a means to "civilize" the native populations, and would provide a physical manifestation of egalitarian idealism (Quinn 1940).

In practice, the township system also meant the newly regimented wild land was looked upon as property—real estate. Every man could own land and alter it for gain, commodifying natural resources. A free, self-reliant landowner would be motivated to establish tenure over a sectionalized acreage, and would establish dominion over the wilderness through his inalienable right of prosperity through toil. A descriptive string of numbers and letters was now available to accurately identify each land parcel—a nomenclature that quantified a myth tied to landscape. Scenes of second-nature landscapes grew to be an encompassing environment for most midwestern inhabitants—establishing the agrarian patterns of the region as a preeminent landscape form that mingled with a democratic American narrative. Many contemporary American observers still so pervasively identify with the Cartesian rationality of the agrarian landscape that the viewshed today remains a precept of Jeffersonian idealism.

Holders of bounty land granted by the federal government after the American War of Independence offered newly measured tracts to newcomers from the East. In Illinois the federal distribution of land amounted to 84 percent of the state's total land area (Bogue 1963, 29). During the early to middle nineteenth century new people continuously came—knowing little of actual conditions, but drawn by the promise of prosperity. Grantees like railroad and land companies, who controlled millions of acres, beckoned disenfranchised easterners with grand visions in periodical advertisements. One ad composed by the Illinois Railroad Company stated, "The attention of persons, whose limited means forbid the purchase of a homestead in the older States, is particularly invited to these lands" (*Harper's Weekly* 1865, 256). Another listing promised "The finest farming lands, equal to any in the world," calling Illinois "The Garden State of America" because "[n]owhere can the industrious farmer secure such immediate results from his labor as on these deep, rich, loamy soils, cultivated with so much ease" (*Harper's Weekly* 1863, 336). Though some found the circumstances on the frontier to be wanting, there were enough successful newcomers to utterly domesticate the environment and to warrant a boom in ventures supporting the great agricultural production.

William Cronon writes, "In Chicago and its hinterland, first and second nature mingled to form a single world. The boosters had been indulging their rhetorical mysticism when they likened the railroads to a force of nature, but there can be no question that the railroads acted as a powerful force *upon* nature, so much so that the logic they expressed in so many intricate ways itself finally came to seem natural" (1991, 93). As railroad companies and storage warehouses became increasingly profitable ventures, the symbiotic relationships between grower, shipper, market, and investor became more difficult to control. Fits and starts in the development of a sophisticated supply chain repeated economic depressions, and the drastic disruptions caused by the Civil War in the early 1860s would lead midwestern agrarians to "foster a sense of regional identity and independence for the Middle West, in part bringing people together and forcing cooperation to temper frontier idealism" (Shortridge 1989, 19). Communities of the West, tied to the land, aligned themselves for the regionally common purpose of self-preservation to survive the competing interests of a perceived eastern capitalist elite.

Midwest Landscape: Manifestation of Values and Activism

As the country's economy grew, the value of the Midwest as a producer grew. A region that had once served as the proving ground for the country's utilitarian expansion from 1800 to 1830 became a strategically important supplier of raw materials and commodities for the nation's livelihood from 1860 to 1890. The region's burgeoning land-based economy rapidly developed in magnitude and complexity, leading to increasingly higher financial stakes, and causing greater market pressure to be exerted on participants at every point of the supply chain. By the end of the nineteenth century these growing pains were accompanied by a more reassuring refrain. The shared narrative outlining the deeds and morality seen as requisite in transforming the Old Northwest Territory from wilderness to landscape had become a touchstone of American values.

During the whole of the nineteenth century, the United States experienced a financial panic nearly every twenty years. Despite a steady increase in agricultural capabilities and expanding farm markets, seemingly external forces like the fluctuating price of gold and availability of currency adversely affected farmers. In the 1870s a majority of the population was engaged in

TABLE 8.1. Chronology of Midwest Political Action

Dates	Party or Movement	Location
1854–	Modern Republican Party	Ripon, Wisconsin; Jackson, Michigan
1861–64	Copperhead Movement	Prominent in Indiana, Illinois, Minnesota, Wisconsin
1867–	The National Grange of the Patrons of Husbandry	Washington, D.C.—but founded by a farmer from Minnesota
1874–84	Greenback Party—Populist Party	Indianapolis, Indiana—Omaha, Nebraska
1897–	Modern Democratic Party (Free Silver factions)	First nominated William Jennings Bryan, Salem, Illinois
1901	Socialist Party of America	Presidential candidacy of Eugene Debs, Terre Haute, Indiana

farming, and popular sentiment felt the panics were caused by a moneyed oligarchy that manipulated markets without restraint. In the Panic of 1869, the price of gold fell from $160 to $133 in fifteen minutes due to the attempt of two brokers to corner the gold market and the subsequent response by the federal government to halt their effort. The fallout from this New York–based financial panic reached midwestern farmers in the form of record low prices for commodities; Chicago grain markets ultimately fell by "a full one-third from a year earlier" (Ackerman 1988, 267).

In a so-called war on wealth, farmers became the vanguard of resistance against the interests of eastern capitalists. Farmers were the prime generators of investment income, and yet the middlemen, such as the proprietors of grain warehouses whose services they required to bring their products to market, squeezed the farmers' profits. As a result, the region's farmers, by means of lobbying through grassroots organizations and movements, sought collective legislative recourse from such circumstances, becoming a strong center of populist activism— ultimately politicizing the landscape in the latter half of the nineteenth century. An examination of the history of American political parties shows one indication of this. Between 1854 and 1901, states in the Midwest became the fountainhead of virtually all major political parties or movements (table 8.1).

Chief among these movements for agrarians was the National Grange of the Patrons of Husbandry (1867). The Grange or Grangers, as they were known, began as a nonpolitical organization founded by an official of the U.S. Department of Agriculture who was concerned with the general state of American agriculture and sought to improve farm family circumstances. The organization's eventual political bent was a continuation of the cause of the so-called Copperheads. This Civil War–era movement of northern Democrats framed the country's mid-century conflict as a struggle between the oppositional interests of eastern capitalists and western producers, rather than the more mainstream understanding of the industrial North versus the agrarian South. The movement's campaign against the Union stemmed from a perceived inequity between eastern markets and western farmers. Lincoln argued for federalist principles to unify a warring country. The Copperheads argued for a regionalist doctrine of western sectionalism that would provide a more economically advantageous position for the Midwest. The Copperheads were "malcontents because they protested against the dwindling influence of agriculture, the nationalization of business, and the centralization of the federal government" (Klement 1969, 324).

The end of the Civil War was the end of the Copperheads, but the issues that fueled their indignation continued to grow. After the Panic of 1873, the Grange expanded rapidly, ultimately establishing twenty thousand

chapters and growing to 1.5 million members. The organization's effective leadership argued for legislation that would ensure its membership could continue to provision the country's needs. As a consequence, the issues of most concern to farmers became key talking points in American political discourse. The financial troubles of the country "quickly became a regional political issue, spawning debates over free silver, protective tariffs, and populist reforms in general" (Shortridge 1989, 19). As a hotbed of aggrieved activists and as an important repository of eastern capital investment, the former territory gained strength as a politically strategic region. Over one twenty-year period, from 1876 to 1896, all major political party conventions, regardless of ideology, were convened in the region (table 8.2; Thompson 1983).

This suggests the great concentration of national interests in the region at the time—whether representing the value of investors' assets or vast labor constituencies. Candidates on the stump readily adopted hyperbolic language of the agrarian narrative for their populist political agendas.

The U.S. Census Bureau uses "the mean center of population" to represent the centroid of the country's population distribution. This term represents an abstract geographic point on which a rigid weightless

TABLE 8.2. Political Party Conventions, 1876–1896

Year	Democratic Conventions	Republican Conventions	Third-party Conventions
1876	St. Louis	Cincinnati	Indianapolis (Greenback)
1880	Cincinnati	Chicago	Chicago (Greenback)
1884	Chicago	Chicago	Indianapolis (Greenback)
1888	St. Louis	Chicago	Cincinnati (Union Labor)
1892	Chicago	Minneapolis	Omaha (People's)
1896	Chicago	St. Louis	St. Louis (People's)

map of the country would balance perfectly when all population members are represented as equal points of mass. According to the Census Bureau, the mean center of population in the forty-eight contiguous United States was located in Ohio, Indiana, or Illinois from about 1855 until about 1975. This means the literal "heartland" of the country for more than one hundred years was somewhere within the Midwest (fig. 8.2). In addition, U.S. Department of Agriculture statistics show that during roughly the same period the percentage of the country's labor force engaged in farming fell dramatically. In the 1850s, farmers constituted 58 percent of the U.S. population, but by the 1970s the proportion had fallen to 4.6 percent (Spielmaker n.d.). The relationship between land and labor had changed dramatically, yet

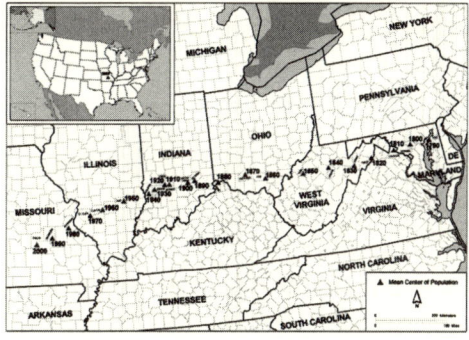

Fig. 8.2. Mean center of population for the United States, 1790–2000. From U.S. Census Bureau, 2000.

an American agrarian narrative where moral and political virtue remain tied to land and labor has persisted.

The period when the region maintained its centralized population coincides with an era of rapid technological innovation and a shift towards manufacturing. In *Middletown: A Study in Modern American Culture*, Robert and Helen Lynd note, "The Federal Census reveals a steady movement westward of the center of manufacturing; in 1880 it was still in Pennsylvania, but by 1890 it had pushed on until it was eight and one-half miles west of Canton, Ohio" (1929, 13). Innovation—both in industry and in agriculture—made possible the workforce's shift from fields to factory floors. Mechanization in agricultural implements meant fewer workers were needed to produce a crop, and mechanization in industry meant less training was necessary to produce goods. A skilled worker in Muncie exclaimed, "But this 'high speed steel' and this new 'stelite' don't absorb the heat and are harder than carbon steel. You can take a boy fresh from the farm and in three days he can manage a machine as well as I can, and I've been at it twenty-seven years" (74).

When the Lynds selected Muncie, Indiana, for their sociological study in the 1920s, they chose a city as representative as possible of contemporary American life. They begin by addressing the two primary eras that made Muncie: the pioneer life in the early part of the century and the discoveries of natural gas that catalyzed industrialization. "Both," they explain, "are within the memory of men who still walk the streets of the city" (10). Those wistful for a "simpler time" remembered ancestral farms and past generations who forged small farms out of wilderness, in part because that was the model familiar to a significant part of the country's population.

As cities and industry drove the country forward into a new century, many cultural commentators expressed suspicion about the shift from agriculture to industrialization that has been well documented in early twentieth-century American literature. Booth Tarkington's *The Turmoil* (1915), Sherwood Anderson's *Winesburg, Ohio* (1919), and Sinclair Lewis's *Main Street* (1920) all chronicled Progressive Era protagonists who negotiate this seemingly incongruent divide. In favoring past or present, Tarkington takes sides on the first page: "There is a midland city in the heart of fair, open country, a dirty and wonderful city nesting dingily in the fog of its own smoke." He continues, "Not so long ago as a generation, there was no panting giant here, no heaving, grimy city; there was but a pleasant big town of neighborly people who had understanding of one another, being on the whole, much of the same type. No one was very rich, few were very poor; the air was clean, and there was time to live" (1915, 9).

The palpable uncertainty about the march of progress expressed in Tarkington's characterizations found a sympathetic audience in midwestern communities and beyond. *The Turmoil* became the number-one bestseller in fiction for the year 1915 (Hackett and Burke 1977). The existential transition from the rural agrarian experience to the industrial urban one found in such fiction relied upon stylized accounts of the past that were understood by mainstream audiences.

Cultural Expediency: Nurturing an American Myth

The marked decline in the farm population during the period from World War I until the Great Depression in 1929 meant that, for the first time since the 1840s, average acreage of farms fell and the total number of farms fell. Mechanization of agricultural tasks and the consolidation of farming interests and national priorities meant that fewer and fewer workers were engaged in farming. It was at this time that the American agricultural mythos, particularly as represented by the Midwest, became a persistent theme in popular media portrayals.

In *The Language of Landscape,* Anne Whiston Spirn describes the deep connection between landscapes and human understanding. She writes, "Landscape, as language, makes thoughts tangible and imagination possible" (1998, 15). In the interest of appealing to the greatest number of viewers, photographers, painters, and filmmakers of the early mass-media era nurtured an American narrative built over the previous century, reinforcing the citizenry's existing sense of the importance of the country's agricultural heritage. Just as late-nineteenth-century activism made landscape the grounds for political struggle, media producers now portrayed the nation's mid-continent landscape as the realm of noble farmers and industrious patriots—codifying American values.

When the Great Depression and a withering drought devastated the country's heartland, the Roosevelt administration consolidated several farm programs under the Resettlement Administration (1935–37) to provide, among other benefits, cash loans to farmers. With the passing of the Farm Security Act in 1937, that agency became the Farm Security Administration (1937–42) and later the Office of War Information (1942–45). Some of the best-known photographers of the twentieth century, including Dorothea Lange, Walker Evans, and Arthur Rothstein, were commissioned to capture the successes of these government programs. Roy Stryker, born in Kansas to a Populist father, was director of this documentary initiative and referred to himself as the "keeper of America's images" (Sandeen 1991, 690). This suggests that Stryker was well aware he was *curating* the American experience—defining rather than just documenting it. The images were framed as documentary but were little more than propaganda for the New Deal programs of the Roosevelt administration. Photographer Ansel Adams quipped that the photographers were "a bunch of sociologists with cameras" (Stryker and Wood 1973, 8).

The agencies collectively recorded more than seventy-seven thousand images over a period of seven years, which were distributed in widely read magazines like *Life* and *Fortune* to promote the legacy of land and the farmer archetype. A select few of these photographs, among the most enduring images ever taken by mid-century photographers, contribute to America's visual heritage and reinforce a constructed agrarian narrative. Lange's *Migrant Mother* (1936) and Rothstein's *Fleeing a Dust Storm* (1936) are still universally recognized as documenting the hardships of 1930s farming and the stalwart perseverance of American agrarians. In fact, the photographers staged both images—in part to garner support for FSA programs and in part because the photographers themselves were steeped in our shared agrarian mythology. They were photographing what they *expected* to find.

Filmmakers would follow similar protocols. *The Plow That Broke the Plains* (1936), commissioned by the Resettlement Administration, is one of the first so-called documentary films produced in the United States. Even though the work is labeled "documentary" in the title sequence, the director, American filmmaker Pare Lorentz, took great license in framing the country's agricultural heritage as a heroic conquest of man against nature. The film also attempts to feature the agricultural industry's contributions to war readiness. The most hyperbolic scenes can be seen about ten minutes into the film. The sequence begins with newspaper headlines such as "England Declares War on Germany" and "War Sends Wheat Soaring." What follows is a thirty-second montage in which farmers on tractors advance from the left in ranks through clouds of dust while German tanks advance from the right through smoke. Periodically the scene cuts to a parade with red, white, and blue bunting overhead to imply a favorable outcome of the skirmish (Resettlement Administration 1936). The film lost its funding due to the fictional treatment of a documentary project, and yet Lorentz felt justified in his account. In his estimate, the power of the agrarian myth was the most relevant of all possible plots.

The degree to which contemporary Americans understand the past through cultural reference should never be underestimated. There are many examples from the 1930s and 1940s in which the expediency of referring to existing cultural artifacts can be seen. The painters known collectively as the Regionalists had a great influence in perpetuating the values of rural circumstances. Thomas Hart Benton (1889–1975), Grant Wood (1891–1942), and John Steuart Curry (1897–1946) were native midwesterners who used a mythological zeitgeist to commemorate agrarian narratives in their works. Due to the popularity of his

1930 painting *American Gothic,* Wood was asked in 1937 to provide illustrations for a Special Editions Club reprinting of Sinclair Lewis's 1920 novel *Main Street* (Delong 2004). The two figures were well matched as cultural producers, even across the span of a decade, because their work was understood by editors, curators, and general audiences to embody the same increasingly nostalgic rural spirit.

John Steuart Curry's *Our Good Earth* (1942) depicts a wheat farmer, discerning and stoic, with broad-brimmed hat and overalls standing in virtual isolation with his wheat crop and children. In that same period the instructional classroom film *The Wheat Farmer* (1938) at one point adopts the same trope and frames the figure in a nearly identical way. The audience relies on their own experiences to perceive the maker's narrative intent—whether that is firsthand experience or experience gathered through consuming media. Referencing iconic cultural portraits to represent a wide range of agrarian circumstances easily transmits a broader, nuanced message to mediated audiences.

Outcomes: Persistence of an Agrarian Landscape Myth

The vestiges of the yeoman farmer and what he embodied can be observed in any number of media scenarios today. The iconography and the mythological agrarian narrative first outlined by boosters in the Old Northwest Territory linger in our collective consciousness. With each presidential campaign race, news media hordes descend on Iowa's second nature to report on the mood of the American people, despite the state's relatively low population. In recent election cycles, even presidents reared in Connecticut are eager to be photographed clearing brush on a ranch, seeking to be portrayed in exactly the manner of the Union's founders, seemingly representing the same populist, agrarian ideals. A recent Super Bowl advertisement for Chrysler Ram Trucks used still images of men working in agricultural landscapes and the voice of the populist radio personality Paul Harvey reading his essay "So God Made a Farmer" to profess their product to be suited to the "farmer in all of us." And the popular television program *Extreme Makeover: Home Edition* still evokes the agrarian spirit found in the tradition of barn-raising. Each episode choreographs shared experiences by eliciting community activism amidst the liberal use of American flags (fig. 8.3).

Much of the United States' cultural heritage is colored by the way in which its system of government came to physically occupy the continent of North America.

Fig. 8.3. Crowd observing the production of *Extreme Makeover: Home Edition* in Philo, Illinois. Photograph by author, 2009.

Manifest Destiny, as divined by God, as orchestrated by its Founding Fathers, and as implemented by its ancestral yeoman farmers, still offers a potent antecedent to its citizenry's relationship to the present-day environment. It is a highly selective narrative—leaving out much that was complicated by conquest, turmoil, or conflict—and therefore as mythological as most origin stories. But for the Midwest region, the narrative is also a parable of an actual landscape, recent and uninterrupted, that was acquired by the United States as a wilderness. The justification for shaping land was necessary for economic or political expediency and provided the means of associating this landscape with democratic idealism, forging the landscape itself into an emblematic American figure over the last two hundred years. The complete transformation of the Old Northwest Territory that began in the early nineteenth century fundamentally contributed to a larger and developing American myth—one that became a consistent and lasting tenet of American culture and values that persists to the present day.

REFERENCES

Ackerman, Kenneth D. 1988. *The Gold Ring: Jim Fisk, Jay Gould, and Black Friday, 1869*. New York: Dodd, Mead & Co.

Bogue, Allan G. 1963. *From Prairie to Corn Belt: Farming on the Illinois and Iowa Prairies in the Nineteenth Century*. Chicago: Quadrangle Books.

Cicero, Marcus Tullius. *De Natura Deorum, Libri Tres*. Trans. Horace C. R. McGregor. Harmondsworth, UK: Penguin, 1972.

Clayton, Andrew R. L. 1996. *Frontier Indiana*. Bloomington: Indiana University Press.

Cronon, William. 1991. *Nature's Metropolis: Chicago and the Great West*. New York: W. W. Norton & Co.

De Tocqueville, Alexis. (1835–40) 2004. *Democracy in America*. Trans. Arthur Goldhammer. New York: Library of America.

DeLong, Lea Rosson. 2004. *Grant Wood's Main Street: Art, Literature, and the American Midwest*. Ames, Ia.: University Art Museums, Iowa State University.

DenUyl, Daniel. 1954. "Indiana's Old Growth Forests." *Proceedings of the Indiana Academy of Science* 63:73–79.

Edwards, Everett E., ed. 1943. *Jefferson and Agriculture: A Sourcebook*. Agricultural History Series No. 7. Washington DC: Bureau of Agricultural Economics, U.S. Department of Agriculture. archive.org/stream/jeffersonagricu107jeff/jeffersonagricu107jeff_djvu.txt (accessed July 30, 2012).

Galbreath, Charles B. 1925. *History of Ohio*. Vol. 1. Chicago: Reprint Services Corp.

Garvey, Timothy J. 1988. *Public Sculptor: Lorado Taft and the Beautification of Chicago*. Urbana: University of Illinois Press.

Hackett, Alice Payne, and James Henry Burke. 1977. *Eighty Years of Best Sellers, 1895–1975*. New York: R. R. Bowker Co.

Hallock, Thomas. 2003. *From the Fallen Tree: Frontier Narratives, Environmental Politics, and the Roots of a National Pastoral, 1749–1826*. Chapel Hill: University of North Carolina Press.

Harper's Weekly. 1863. "The Finest Farming Lands (Equal to Any in the World!!!)." Advertisement (May 23). www.sonofthesouth.net/leefoundation/civil-war/1863/may/fernando-wood-cartoon.htm (accessed July 30, 2012).

———. 1865. "Best Farming Lands in the World (for Sale by the Illinois Central Railroad Co.)." Advertisement (April 22). www.sonofthesouth.net/leefoundation/civil-war/1865/April/illinois-central-railroad-land.htm (accessed July 30, 2012).

Hart, John Fraser. 1968. "Field Patterns in Indiana." *Geographical Review* 58:450–71.

Havighurst, Walter. 1946. *Land of Promise: The Story of the Northwest Territory*. New York: Macmillan Co.

———. 1956. *Wilderness for Sale: The Story of the First Western Land Rush*. New York: Hastings House.

Klement, Frank L. 1969. "Middle Western Copperheadism and the Genesis of the Granger Movement." In *The Old Northwest, Studies in Regional History, 1787–1910*, ed. Harry N. Scheiber, 323–40. Lincoln: University of Nebraska Press.

Lynd, Robert S., and Helen Merrell Lynd. 1929. *Middletown: A Study in Modern American Culture*. New York: Harcourt Brace & Co.

Nash, Roderick. 1967. *Wilderness and the American Mind*. New Haven: Yale University Press.

Nye, David E. 1994. *American Technological Sublime*. Cambridge, MA: MIT Press.

Quinn, Patrick F. 1940. "Agrarianism and the Jeffersonian Philosophy." *Review of Politics* 2 (January): 87–104.

Resettlement Administration (producer), and Pare Lorentz (director). 1936. *The Plow That Broke the Plains*. Film. San Francisco: Prelinger Internet Archives, Prelinger Library. archive.org/details/PlowThatBrokethePlains1 (accessed July 30, 2012).

Rohrbough, Malcolm J. 2008. *Trans-Appalachian Frontier: People, Societies, and Institutions, 1775–*

1850. 3rd ed. Bloomington: Indiana University Press.

Sandeen, Eric J. 1991. "The Search for Meaning in Farm Security Administration Photographs." *American Quarterly* 43 (December): 688–93.

Scharf, Albert F. 1900. *Indian Trails and Villages of Chicago, and of Cook, Dupage and Will Counties, Illinois, 1804.* Map. www.franzosenbuschheritage project.org/Histories/YorkTownshipMaps/Native American%20Settlements%20of%201804.htm (accessed July 30, 2012).

Scott, David B. 1884. "Washington's Journey." In *A School History of the United States.* New York: American Book Co. etc.usf.edu/ clipart/30400/30415/wash_journ_30415.htm# .UKFDcY6fOq8 (accessed July 30, 2012).

Shortridge, James R. 1989. *The Middle West: Its Meaning in American Culture.* Lawrence: University Press of Kansas. Smith, Henry Nash. 1950. *Virgin Land.* Cambridge, MA: Harvard University Press.

Sommers, Lawrence M., ed. 1977. *Atlas of Michigan.* www.geo.msu.edu/geomich/Treaties.html (accessed July 30, 2012).

Spielmaker, Debra. N.d. *Growing a Nation: The Story of American Agriculture, Historical Timeline— Farmers and the Land.* Utah State University Extension. www.agclassroom.org/gan/timeline/ farmers_land.htm (accessed July 30, 2012).

Spirn, Anne Whiston. 1998. *The Language of Landscape.* New Haven, CT: Yale University Press.

Stilgoe, John R. 1982. *Common Landscape of America, 1580 to 1845.* New Haven, CT: Yale University Press.

Stryker, Roy Emerson, and Nancy C. Wood. 1973. *In This Proud Land: America, 1935–1943, as Seen in the FSA Photographs.* Greenwich, CT: New York Graphic Society.

Tarkington, Booth. 1915. *The Turmoil.* New York: Grosset & Dunlap.

Thompson, Margaret C., ed. 1983. *Presidential Elections since 1789.* Washington, DC: CQ Roll Call.

U.S. Census Bureau. 2000. *Mean Center of Population for the United States, 1790 to 2000.* Washington DC: U.S. Department of Commerce, Economics and Statistics Administration, Geography Division. www.census.gov/geo/www/cenpop/meanctr.pdf (accessed July 30, 2012).

"Washington as a Farmer." 1855. Engraving. New York Public Library Mid-Manhattan Library Picture Collection Online. catalog.nypl.org/search~S1?/ .b17180963/.b17180963/1,1,1,B/1856~b17180963 &FF=&1,0,,1,0 (accessed July 30, 2012).

Watts, May Theilgaard. 1975. *Reading the Landscape of America.* New York: Macmillan.

Weeks, William Earl. 1996. *Building the Continental Empire: American Expansion from the Revolution to the Civil War.* Chicago: Ivan R. Dee.

JENNIFER D. W. BRITTON

Imbibing *Terroir*

VALUES IN NAPA VALLEY'S CULTURAL LANDSCAPE OF WINE

As a social construction of nature, viticulture has evolved as one of the earliest and most enduring built environments (McGovern 2003); devoted to the production of wine, these landscapes exhibit some of the most complex and enigmatic interchanges of human values. In global recognition of the ideological, economic, political, and cultural interactions of these landscapes with the environment, the United Nations Educational, Scientific and Cultural Organization (UNESCO) has designated eight viticultural landscapes as World Heritage Cultural Sites, with an additional six under consideration as of this writing (UNESCO 2012). In the belief that the landscape of wine and vine reflects a

society (Cosgrove 1998; Murphy 2005), this interpretive study explores the way we ascribe the values of pleasure, paradise, and power to the cultural landscape and how these symbolic layers enable consumers to transcend reality in pursuit of an ideal place.

Although people widely appreciate wine landscapes, research to date has seldom explored *how* and *why* we value it (Dougherty 2011; Peters 1997; Schivelbusch 1992; Sommers 2008; Ulin 1996; Unwin 1996). Instead, most inquiry has focused on the science, technology, and history of cultivation (Amerine and Singleton 1965; Boulton 1996; Lawrence 1989; Fleming 2001; McGovern

2003; Pinney 1989, 2005; Sullivan 1994; Winkler 1974; Zoecklein 1995). Yet this unique growing environment—*terroir*—is brought into being through something more than the "physical aspects of geology, soils, climate, geomorphology and vegetation" (Unwin 2011, 37). Because the wine landscape embodies a place beyond physical utility, an assay may benefit individuals who critically shape, create, plan, manipulate, and reenvision landscape's aesthetic qualities (American Society of Landscape Architects 2012; American Planning Association 2009). To these ends, professionals typically take cues from history and theory, engage in open and ongoing public debate, gather site inventories and analyses, and participate with allied professions. However, to address societal requirements, assuming "landscape is composed not only of what lies before our eyes but what lies within our heads" (Meinig 1979, 34), we must be mindful of the "bond between people and place" (Tuan 1990, 4). Otherwise, little chance exists of ensuring improvements in the human condition as we are more likely to merely create temporary alterations to visual form or style. Investigating our complex relationship with the significant cultural landscapes of viticulture therefore provides an opportunity for greater insight into creating places of lasting value.

Methods

The scope of this interpretation focuses on a well-known and significant American viticultural landscape—Napa Valley, California. In order to investigate Napa's wine landscape in context, with reasonable textural depth and verisimilitude, this research study triangulated: (1) a review of historical development, (2) a personal phenomenological descriptive aesthetic (Berleant 1992), and (3) a semiotic analysis of an advertisement as an important "ideological form(s) in contemporary capitalist societies" (Rose 2001, 70).

The data gathered included: archival research into viticulture's geology, climate, economics, and politics in America and specifically the Napa Valley; field recordings of direct, unobtrusive observation from naturally occurring events on Thursday, February 9, 2006, along with written reflection of elements touched, smelled, tasted, seen, and remembered; and a sampling of magazine advertisements to analyze consumer motivation or means-end chain (Charters 2006, 136) in decisions guiding consumption.

To choose an advertisement for analysis, a preliminary data set was recorded from three lifestyle magazines, *Travel, Food & Wine,* and *Home & Interior.* This sample was cross-referenced with a 2005 calendar-year collection of *Gourmet, Food & Wine, House Beautiful,* and *Real Simple.* From these publications all wine advertisements pertaining to the Napa Valley were divided into six communicative themes (Rose 2001): socialization, food, escape, sense of place, sophistication/elite, and value/quality. Escape emerged as the overarching dominant theme, with a Beringer advertisement featuring Chicago repeating most often (fig. 9.1). This advertisement, with strong patterns in meaning exchange, was subsequently used for semiotic analysis.

To reveal the advertisement's deeper messages, semiotic analysis identified orders of signification, those individual images and text within the whole that communicated unique meanings: (1) Chicago aerial, (2) vineyard, (3) Beringer wine bottle, (4) textual directions, (5) Beringer 2004 copyright, and (6) tag line. The wine-label subtext included: (7) varietal (Chardonnay), (8) vintage year (2004), and (9) Beringer Rhine House image. This evidence was examined for both literal (denotative) meaning and the message's effect (connotation) on the viewer in the context of the advertisement (Sebeok 2001; Chandler 2002, 137). Finally, to determine how *value* may be assigned to specific imagery, this research employed repeated commutation tests, an analytical

technique that determines the strength of the signifiers by replacing images and/or text to expose paradigms, and to determine "whether a change on the level of the signifier leads to a change on the level of the signified" (Chandler 2002, 89).

The final interpretation resulted from identifying and triangulating—in a circuitous process of "mapping, exposure, and intertextuality" (Schwartz-Shea and Yanow 2012, 87)—those culturally perceived values in Napa Valley's wine landscape that were consistently manifested across all data sources.

Results and Discussion

Viewing this landscape through historical, observed, and visual culture revealed three dominant symbolic dimensions that enable landscape appreciation to eclipse realities: (1) hedonic pleasure of the body and mind, (2) mythical paradise based in religion and American ideology, and (3) social power linked to viewing land as commodity.

HEDONIC PLEASURE

Historically, many cultures have lauded wine for physical utilitarian benefits (Charters 2006), and a successful viticultural landscape

Fig. 9.1. Wine advertisement used for semiotic analysis. Used with permission of Beringer Vineyards.

How to get to **NAPA VALLEY, CA,** from **CHICAGO, IL.**

Start at **W. ARMITAGE AVE** heading WEST   go **0.3 mile**
Turn RIGHT on to **N. CLYBOURN AVE**   go **0.4 mile**
Turn RIGHT into **DIRK'S FISH & GOURMET SHOP   park car**
Walk NORTH   **0.0001 mile**
Take **SALMON   8 ounces**
Take **CALAMARI   4 ounces**
Take **SHRIMP   9**
Take **MUSSELS   8**
Take **TOMATO SAFFRON BROTH   12 ounces**
Continue toward checkout   go **0.00015 mile**
Merge into traffic and drive until you reach **HOME**
Prepare ingredients for **SEAFOOD STEW   0.3 hour**
Merge **CORKSCREW** with bottle of **BERINGER**
Pour **BERINGER** into glass   **4 ounces**
Serve **SEAFOOD STEW   large bowl**
Sip **BERINGER** and arrive in **NAPA VALLEY, CA**
Total travel time   **45 minutes**

BERINGER - how to get to napa valley

offered the pleasure of a more healthful existence. Humans have used wine as a remedy for typhoid fever, as a gentle narcotic (Lawrence 1989), and in prevention of cardiovascular disease (Lucia 1954; Lewin 2010). Proverbial wisdom in the New Testament (for example 1 Timothy 5:23) describes wine as preferable to water for medicinal use (Carroll, ed. 1997) and Christian monks, working at the forefront of medical and botanical knowledge, planted vineyards in Europe and the New World (Lewin 2010) in part as curative resources for various maladies (Fuller 1996; Harvey 1981). As evident today, the word "Mönch" or "Moines" (meaning monk) in a wine's name refers to its monastic roots (Hales 2000). Likewise, the Christian Brothers namesake, replete through Napa Valley history, speaks to the legacy of Napa's Spanish Franciscan monastic wine growing (Hales 2000). Californian missions and American colonists struggled to achieve stable wine production with both the European varietal *Vitis vinifera* and wild North American grapes in their desire for a reliable and safe beverage (Mendelson 2009; Pinney 1989). Thomas Jefferson planted vineyards at his Monticello home in the belief that the nation's health depended on viticultural development (Lawrence 1989).

Beyond its salubrious properties, cultivation of the wine landscape expresses a shift in human agricultural pursuits as "with the advent of viticulture, agriculture turned toward producing a product not necessary for biological subsistence, but still important on a *social* level, in that wine brought enjoyment" (Walsh 2000, 11–12). While cereal grains provided stable nourishment, as early as 7000 B.C.E. to 5400 B.C.E. (McGovern 2003, Hales 2000, Lewin 2010) cultivated vineyards provided long-term storable, tradable pleasure. A Roman epitaph in the comprehensive collection of ancient Latin inscriptions, the *Corpus Inscriptionum Latinarum*, reads, "Baths, wine, and sex ruins our bodies. But what makes life worthwhile except baths, wine and sex?" (Fleming 2001, 1). Vine and wine afforded sensual, even aphrodisiac pleasures, both from taste and alcohol stimulation (Broshi 2001, 164–65). Wine's psychotropic ability to reduce inhibitions and foster feelings of well-being thus links viticulture with social solidarity, interpersonal warmth and intimacy, family meals, and other shared human values (Charters 2006, Fleming 2001, McGovern 2003).

Similarly, visitors to Napa Valley may experience these satiating, restorative qualities in the landscape's tapestry of wineries, vineyards, and restaurants removed from urban pressures, whether of San Francisco's Bay Area or Chicago's metropolis

Fig. 9.2. V. Sattui Winery, offering idyllic picnic grounds, deli, and marketplace set among the vineyards and winery. Photo by author, 2006.

Fig. 9.3. Enlargement of Beringer advertisement's "directional recipe." Used with permission of Beringer Vineyards.

How to get to NAPA VALLEY, CA, from CHICAGO, IL.

Start at **W. ARMITAGE AVE** heading WEST go **0.3 mile**
Turn RIGHT on to **N. CLYBOURN AVE** go **0.4 mile**
Turn RIGHT into **DIRK'S FISH & GOURMET SHOP** park car
Walk NORTH **0.0001 mile**
Take **SALMON** **8 ounces**
Take **CALAMARI** **4 ounces**
Take **SHRIMP** **9**
Take **MUSSELS** **8**
Take **TOMATO SAFFRON BROTH** **12 ounces**
Continue toward checkout go **0.00015 mile**
Merge into traffic and drive until you reach **HOME**
Prepare ingredients for **SEAFOOD STEW** **0.3 hour**
Merge **CORKSCREW** with bottle of **BERINGER**
Pour **BERINGER** into glass **4 ounces**
Serve **SEAFOOD STEW** **large bowl**
Sip **BERINGER** and arrive in **NAPA VALLEY, CA**
Total travel time **45 minutes**

(fig. 9.2). Beringer's advertisement includes the notion of pleasure because it denotes imbibing with good food (fig. 9.3), while it connotes an antithesis in the perceived contrasts of environmental health between downtown Chicago and a Napa vineyard. Depicted as a sepia aerial image (fig. 9.1), Chicago dominates the advertisement, simultaneously distancing the viewer while implying escape through flight. In contrast, lower on the same page, an eye-level vineyard image rich in color visually indicates Napa's *terroir*, a bountiful, healthful landscape, an embodiment of pleasure, as the ultimate escape destination.

Yet, in reality, is this comparison fair? Although environmentally sustainable grape production has become more prevalent (Hall and Mitchell 2008, 92–109), according to Napa County's 2011 Farm Report only 8.1 percent of Napa's fruit-bearing vineyards use organic farming methods (County of Napa 2011). In 2003 Napa County used 1,455,190 gross pounds of pesticides, and of the top fifty pesticides used in grape production, eleven were considered "PAN Bad Actor Pesticides" or most toxic pesticides (Kegeley et al. 2011). Also, as production pressure has increased, bare ridgelines from hasty hillside development affect watershed quality, resulting in a landscape resembling an industrial rather than an agricultural artifact (Conaway 1990). During the 1990s, district staff measured loss of topsoil in several vineyards with "as much as fourteen tons per acre each year from runoff" (Sullivan 1994, 354). Erosion from hillside development became so environmentally disastrous that in November 1989 the *Star* (a local newspaper) called for measures to save Napa hillsides and their topsoil (Sullivan 1994).

Regardless of this environmental reality, the purpose of Beringer's advertisement is to sell a product through a message of physical and mental pleasure associated with Napa's wine landscape. By utilizing the image of Beringer's bottle, full of hedonistic elixir,

as the connotative mechanism for artificial transport to Napa's viticultural landscape (seen next to it), the advertisement capitalizes on imagined physical satisfaction in a pleasurable place. This message becomes apparent in a commutation test by replacing the vineyard imagery with diesel smudge pots (fig. 9.4), a common device used for frost protection. When examined in this context, the illusion of environmental quality in the advertisement dissolves—immediately casting the viewer back into Chicago's dense urban fabric. The lure of Napa's *terroir* loses aesthetic potency for escape, while in comparison the aerial view of Chicago's Magnificent Mile becomes the more physically pleasing landscape.

MYTHICAL PARADISE

For early Christians, viticulture and wine were considered gifts from God and symbolized a good omen of peace and tranquility (Hales 2000). In settling the New World, Europeans attempted to reconstruct a new earth resembling the original idyllic Edenic garden (Moynihan 1979; Schivelbusch 1992), and Thomas Jefferson promoted viticulture as an agrarian democratic expression (Lawrence 1989). As a place recovered from unpredictable wilderness, American viticulture thus represents the

Fig. 9.4. Commutation test: technology. Beringer advertisement modified with photograph of smudge pots added. Photomontage by author, 2006.

How to get to **NAPA VALLEY, CA,** from **CHICAGO, IL.**

Start at **W. ARMITAGE AVE** heading WEST   go **0.3 mile**
Turn RIGHT on to **N. CLYBOURN AVE**   go **0.4 mile**
Turn RIGHT into **DIRK'S FISH & GOURMET SHOP**   park car
Walk NORTH   **0.0001 mile**
Take **SALMON**   **8 ounces**
Take **CALAMARI**   **4 ounces**
Take **SHRIMP**   **9**
Take **MUSSELS**   **8**
Take **TOMATO SAFFRON BROTH**   **12 ounces**
Continue toward checkout   go **0.00015 mile**
Merge into traffic and drive until you reach **HOME**
Prepare ingredients for **SEAFOOD STEW**   **0.3 hour**
Merge **CORKSCREW** with bottle of **BERINGER**
Pour **BERINGER** into glass   **4 ounces**
Serve **SEAFOOD STEW**   **large bowl**
Sip **BERINGER** and arrive in **NAPA VALLEY, CA**
Total travel time   **45 minutes**

BERINGER - how to get to napa valley

union between divine and human activities where society and earth join in perfection (Broshi 2001; Conaway 1990).

Napa Valley's wine landscape displays this transformation of nature into ordered, cultivated civil society. The unique *terroir* is fully and legally defined by human process with sixteen American Viticulture Areas (AVAs) and including 45,420 total cultivated acres (County of Napa 2011). The landscape exists as a tamed agrarian place devoid of grizzly bears, Wappo Indians, oaks, and manzanita; in their stead stand ordered vines, trellised to wires and controlled by humans.

Unlike the Beringer imagery of a fertile Napa Valley bereft of laborers, in reality this landscape exists as an artifact of unrelenting hard work (Daniel 1981; Goldfarb 1981), and wine industrialists have relied heavily, if not exclusively, on a steady, affordable immigrant labor force (Daniel 1981). The Beringer advertisement does not directly communicate whose labors modify or maintain the environment, neither the nineteenth-century immigrant Chinese workers who built Beringer Vineyards nor the predominantly Latino laborers who work in the vineyard today. Beringer's website only attributes Chinese labor with digging the wine tunnels, and no mention appears regarding today's laborers (Beringer Vineyards 2013).

The omission of immigrant farmworkers in wine advertisements is historically commonplace. In the 1880s, *Harper's Weekly* published an advertisement depicting Chinese immigrant men stomping grapes with their bare feet. At the time, the advertisement infuriated vintners as they felt some inaccuracies, such as stomping grapes, damaged the reputation of Napa wineries. Further, they "demanded the magazine drop all further reference to the Chinese" to avoid "linking their product to Chinese hands" (Street 2004, 317–18).

This tenuous relationship between advertising message and the reality of labor becomes noticeable by replacing the flawless vineyard imagery with direct reference to laborers "adrift in a landscape of ordered beauty" (Street 2004, xviii). As seen in the commutations test, without the vineyard photograph as a mediator of style and a way of life (Ewen 1988), the perception of paradise erodes (fig. 9.5). Based in the desire to sell a commodity, Beringer's lacquered reality is not unique to wine, but unlike an advertisement for blue jeans that may appropriate the western landscape for product mythos, or a tourism advertisement that may promote the landscape itself as a commodity, Beringer advertises their wine's desirability relative to the quality of the place of production (Napa Valley's wine landscape). Thus portraying the human cost required to

Fig. 9.5. Commutation
test: labor. Beringer
advertisement modified
with photograph of
field laborers added.
Photomontage by author,
2006.

How to get to **NAPA VALLEY, CA,** from **CHICAGO, IL.**

Start at **W. ARMITAGE AVE** heading WEST   go **0.3 mile**
Turn RIGHT on to **N. CLYBOURN AVE**   go **0.4 mile**
Turn RIGHT into **DIRK'S FISH & GOURMET SHOP**   park car
Walk NORTH   **0.0001 mile**
Take **SALMON   8 ounces**
Take **CALAMARI   4 ounces**
Take **SHRIMP   9**
Take **MUSSELS   8**
Take **TOMATO SAFFRON BROTH   12 ounces**
Continue toward checkout   go **0.00015 mile**
Merge into traffic and drive until you reach **HOME**
Prepare ingredients for **SEAFOOD STEW   0.3 hour**
Merge **CORKSCREW** with bottle of **BERINGER**
Pour **BERINGER** into glass   **4 ounces**
Serve **SEAFOOD STEW   large bowl**
Sip **BERINGER** and arrive in **NAPA VALLEY, CA**
Total travel time   **45 minutes**

BERINGER - how to get to napa valley

produce Napa's iconic landscape drains the mythic and pastoral paradise, and ultimately negates the advertisement's ability to evoke escape.

SOCIAL POWER

Land as an object to be purchased and, in effect, consumed has long been a symbol for social status (Mitchell 1994); because of its worth as an exchange commodity, the viticultural landscape has shaped society's hierarchies of power. Neolithic upper classes defined their status, resources, and leisure through planting vineyards and enjoying wine (McGovern 2003; Unwin 1996; Heath 2000). For the Egyptians and Mesopotamians, obtaining wine was restricted to the wealthy and those wielding imperial power (Walsh 2000; Lewin 2010). During Greek and Roman antiquity, wine and its consumption were "held in highest regard" (Wilkins and Hill 2006, 170) with the best quality reserved for upper classes. Celtic society saw wine as "a prestige commodity—a sign of wealth ranking alongside gold and thus an instrument of power" (Fleming 2001, 5).

Early American colonists aspired and struggled for wine production in search of a prosperous life, for they identified wine and viticulture with a culturally sophisticated and successful society. Thomas Jefferson believed wine indicated refinement and associated viticulture with high culture, because connoisseurship required discernment (Fuller 1996). Yet America has had an historically fickle relationship with wine. The viticultural landscape, as the producer of an alcoholic product, has been guilty by association with drunkenness, moral weakness, and lawlessness. The "noble experiment" of Prohibition pendulated cultural opinion from a "demon to darling" status (Mendelson 2009). Since Prohibition's repeal, and in particular since the 1970s, America has witnessed a significant shift in wine culture with a large concentration of consumers within college-educated professional classes (Moulton and Lapsley 2001; Mendelson 2009). As a result, disproportionate wealth has settled in Napa Valley with emblematic "social markers" that denote economic class (figs. 9.6 and 9.7).

Aesthetic appreciation of Napa's viticultural landscape clearly involves class-based dimensions. Similar to collecting fine art, owning a portion of Napa Valley's wine landscape conveys a sign of distinction (Conaway 2002). As the wine and viticulture historian Charles Sullivan pointed out, "An economist for Bank of America put forth a view that was as sound in 1885 as it was in 1980. 'The Napa Valley is a unique commodity that demands a premium, like buying a unique painting. It's a unique socio-economic environment that takes a certain

Fig. 9.6. Private home
in Napa Valley. Photo by
author, 2006.

Fig. 9.7. Upon entry to Napa Valley, a bronze statue depicting the crushing of grapes. Photo by author, 2006.

clientele'" (1994, 322). For those privileged enough to own a vineyard, Napa's wine landscape acts much like a fine bottle of wine, "as a social marker or a sign of belonging to a dominant social class" (Demossier 2005, 133).

By strategically advertising in specific lifestyle magazines, Beringer targets a consumer who desires social status. This class dimension becomes most apparent in the connotative and denotative messaging addressed in the text: to "arrive" in the desired Napa landscape requires ingredients purchased from Dirk's Fish & Gourmet Shop, reputed for expensive, fresh seafood and located in Lincoln Park—one of Chicago's most affluent areas with four-star restaurants and stylish shopping (fig. 9.3). The advertisement then supplies symbolic contrasting imagery of Napa's wine landscape juxtaposed to an abstracted portrayal of Chicago. The textural description provides a non-inclusive connotative message to "transport" or escape from the mundane Chicago existence for an ideal landscape, but a consumer can only accomplish this through purchasing a fifteen-to-forty-dollar bottle of wine, salmon, calamari, shrimp, and mussels and cooking a tomato saffron broth. Thus the metaphoric escape to Napa's wine landscape becomes available only for those with disposable income and leisure.

In reality, Napa's local landscape also participates in a larger, global hierarchy of wealth and power. In the Wine Institute's 2011 wine sales statistical report, California "held a 61 percent share of the U.S. market"

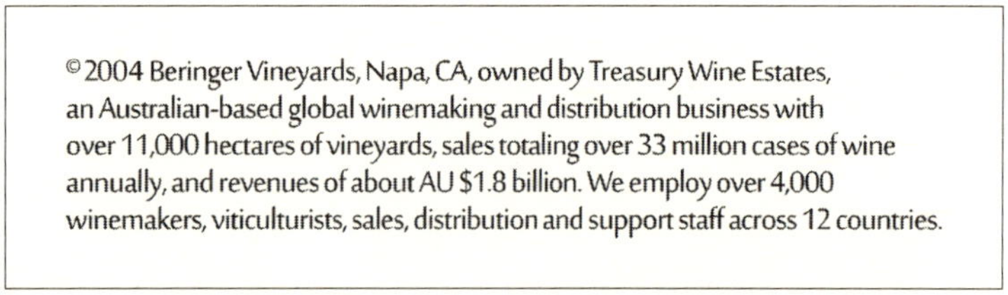

Fig. 9.8. Commutation test: global capital. Beringer advertisement modified with indication of corporate globalization. Photo manipulation by author.

(Wine Institute 2012), with Napa's grape and wine industry contributing a significant "21% of the total economic impact of wine in California" (County of Napa 2009). With an estimated $9.5 billion from Napa's wine and viticulture sector, profitability fosters a globalized corporate presence in the valley; two out of the four biggest landowners are controlled by interests outside of California (Winter 2001). Napa's wine landscape functions as an "extension of agribusiness" (Tyrrell 1999, 14) controlled by beverage conglomerations in wealthy produce exchanges. In the case of Beringer Vineyards, the oldest continuously operating winery in Napa Valley, the winery and vineyards were sold to Nestlé in 1971, Foster's Group in 1996, and are now an asset in Treasury Wine Estates' wine portfolio (Treasury Wine Estates 2012). A family winery name such as "Beringer," once associated with a particular place, *terroir*, and product, is today simply a label in a corporate portfolio. However, because a belief or image of established cultural and natural heritage enables emotive consumption (Conaway

1990; Hall and Mitchell 2008; Lewin 2010), Beringer's advertisement focuses on place-based authenticity and *terroir*. So, too, is this message reinforced on their website with repeating headlines: "Beringer is built on Heritage . . . is rooted in Napa Valley . . . is Napa Valley" (Beringer Vineyards 2013). However, as demonstrated in a final commutation test (fig. 9.8), replacing Beringer's trademark with a more accurate description of its global corporate ownership interrupts the aesthetic illusion of escape and participation in the types of social power extended by drinking fine wine.

Final Reflection

Our pale day is sinking into twilight,
And if we sip the wine, we find dreams
   coming upon us
Out of the imminent night.
       —D. H. Lawrence (1992)

The cultural motivations to create and maintain Napa's viticultural landscape stem from its ability to detach itself from the realities of its creation and sustainment, to transcend issues of environment, labor, and global capital, and to provide us with desirable physical and imaginative reverie (Smith 1993). As the consumption of Napa's wine grows worldwide (Wine Institute

2012), so too has Napa's wine landscape become alluring in promise to "lift us out of the dreariness of necessity" (Ewen 1988, 14). During the course of performing field research, I observed a place manicured by money. Wineries and estate homes stood emblematically among well-groomed vineyards, boutique shops, and gourmet restaurants. People sat at leisure in cafés during the middle of a workday. The environment felt surreal and sheltered—a desirable oasis from urban life.

Although human beings have only three basic needs (food, shelter, and clothing), to flourish we also seek "places which afford us pleasure and mental stimulation; environments that supply an indication of our past and of what the future might hold" (Gold and Burgess 1982, 1). In order to address societal requirements and create settings for human activity, designers and planners must be challenged to consider values in landscapes, the dynamic interaction between people and environment, contingent on convention, human perception, and social experience. In interpreting the land of wine and vine, we behold the heart of a culture. Viticulture transfigures the landscape through mankind's shaping intention to create an ideal world. There is a saying in Latin: *in vino veritas*—truth is in the wine. Yet verity also exists in the mind, body, and soul of a wine landscape and, like a fine

vintage, the complex quality of pleasure, paradise, and power in Napa's viticultural landscape creates a unique and lasting place.

REFERENCES

American Planning Association. 2009. "Code of Ethics and Professional Conduct." www.planning.org/ethics/ethicscode.htm (accessed November 14, 2012).

American Society of Landscape Architects. 2012. "Mission Statement." www.asla.org/MissionStatement.aspx (accessed January 6, 2012).

Amerine, Maynard A., and Vernon L. Singleton. 1965. *Wine, an Introduction for Americans.* Berkeley: University of California Press.

Beringer Vineyards. 2005. "How to Get to Napa Valley, Ca, from Chicago, Il." Advertisement. *Wine Spectator* 30(3): 33. May 31.

———. 2013. "Heritage." www.beringer.com/ (accessed November 14, 2012).

Berleant, Arnold. 1992. *The Aesthetics of Environment.* Philadelphia: Temple University.

Boulton, Roger B. 1996. *Principles and Practices of Winemaking.* New York: Chapman & Hall.

Broshi, Magen. 2001. *Bread, Wine, Walls and Scrolls.* Sheffield, UK: Sheffield Academic Press.

Carroll, Robert Prickett Stephen, ed. *The Bible: Authorized King James Version.* 1997 edition. Oxford, UK: Oxford University Press.

Chandler, Daniel. 2002. *Semiotics: The Basics.* London: Routledge.

Charters, Stephen. 2006. *Wine and Society: The Social and Cultural Context of a Drink.* Amsterdam: Elsevier/Butterworth-Heinemann.

Conaway, James. 1990. *Napa.* New York: Avon Books.

———. 2002. *The Far Side of Eden: New Money, Old Land, and the Battle for Napa Valley.* Boston: Houghton Mifflin.

Cosgrove, Denis E. 1998. *Social Formation and Symbolic Landscape*. Madison: University of Wisconsin Press.

County of Napa. 2009. "We Are Napa County Public Information: Grapes and Wine Industry." www .countyofnapa.org/Pages/DepartmentContent .aspx?id=4294967635 (accessed January 21, 2012).

———. 2011. *The Napa County Department of Agriculture and Weights and Measures 2011 Agricultural Crop Report*. www.countyofnapa.org/ WorkArea/DownloadAsset.aspx?id=4294976456 (PDF version of document downloaded November 9, 2012).

Daniel, Cletus E. 1981. *Bitter Harvest: A History of California Farmworkers, 1870–1941*. Berkeley: University of California Press.

Demossier, M. 2005. "Consuming Wine in France: The 'Wandering' Drinker and the Vin-Anomie." In *Drinking Cultures: Alcohol and Identity*, ed. Thomas M. Wilson. Oxford, UK: Berg Publishers.

Dougherty, Percy H. 2011. *Viticulture: The Geography of Wine*. Dordrecht, Netherlands: Springer.

Ewen, Stuart. 1988. *All Consuming Images: The Politics of Style in Contemporary Culture*. New York: Basic Books.

Fleming, S. J. 2001. *Vinum: The Story of Roman Wine*. Glen Mills, PA: Art Flair.

Fuller, Robert C. 1996. *Religion and Wine: A Cultural History of Wine Drinking in the United States*. Knoxville: University of Tennessee Press.

Gold, John Robert, and Jacquelin A. Burgess. 1982. *Valued Environments*. London: G. Allen & Unwin.

Goldfarb, Ronald L. 1981. *Migrant Farm Workers: A Caste of Despair*. Ames: Iowa State University Press.

Hales, Michael. 2000. *Monastic Gardens*. New York: Stewart, Tabori & Chang. catalog.hathitrust.org/api/ volumes/oclc/43333702.html.

Hall, Colin Michael, and Richard Mitchell. 2008. *Wine Marketing: A Practical Guide*. Amsterdam: Elsevier.

Harvey, John. 1981. *Mediaeval Gardens*. Beaverton, OR: Timber Press

Heath, Dwight B. 2000. *Drinking Occasions: Comparative Perspectives on Alcohol and Culture*. Philadelphia: Brunner/Mazel.

Kegeley, S. E., B. R. Hill, S. Orme, and A. H. Choi. 2011. *Pan Pesticides Database*, Pesticide Action Network, North America. San Francisco. www .pesticideinfo.org (accessed January 20, 2012).

Lawrence, D. H. 1992. *Birds, Beasts and Flowers: Poems*. Santa Rosa, CA: Black Sparrow Press.

Lawrence, R. de Treville, III. 1989. *Jefferson and Wine: Model of Moderation*. The Plains, VA: Vinifera Wine Growers Association.

Lewin, Benjamin. 2010. *Wine Myths and Reality*. Dover, UK: Vendange Press.

Lucia, S. P. 1954. *Wine as Food and Medicine*. New York: Blakiston Co., Inc.

McGovern, Patrick E. 2003. *Ancient Wine: The Search for the Origins of Viniculture*. Princeton, NJ: Princeton University Press.

Meinig, D. W. 1979. "The Beholding Eye: Ten Versions of the Same Scene." In *The Interpretation of Ordinary Landscapes: Geographical Essays*, ed. D. W. Meinig. New York: Oxford University Press, Inc.

Mendelson, Richard. 2009. *From Demon to Darling: A Legal History of Wine in America*. Berkeley: University of California Press.

Mitchell, W. J. T., ed. 1994. *Landscape and Power*. Chicago: University of Chicago Press.

Moulton, Kirby S., and James T. Lapsley, eds. 2001. *Successful Wine Marketing*. Gaithersburg, MD: Aspen Publishers.

Moynihan, Elizabeth B. 1979. *Paradise as a Garden: In Persia and Mughal India*. World Landscape Art & Architecture Series. New York: G. Braziller.

Murphy, Michael D. 2005. *Landscape Architecture Theory: An Evolving Body of Thought*. Long Grove, IL: Waveland Press, Inc.

Peters, Gary L. 1997. *American Winescapes: The Cultural Landscapes of America's Wine Country*. Boulder, CO: Westview Press.

Pinney, Thomas. 1989. *A History of Wine in America: From the Beginnings to Prohibition*. Berkeley: University of California Press.

———. 2005. *A History of Wine in America: From Prohibition to the Present*. Berkeley: University of California Press.

Rose, Gillian. 2001. *Visual Methodologies: An Introduction to the Interpretation of Visual Materials*. London: Sage.

Schivelbusch, Wolfgang. 1992. *Tastes of Paradise: A Social History of Spices, Stimulants, and Intoxicants*. New York: Pantheon Books.

Schwartz-Shea, Peregrine, and Dvora Yanow. 2012. *Interpretive Research Design: Concepts and Processes*. New York: Routledge.

Sebeok, Thomas A. 2001. *Signs: An Introduction to Semiotics,* 2nd ed. Toronto: University of Toronto Press.

Smith, Jonathan. 1993. "The Lie That Blinds: Destabilizing the Text of Landscape." In *Place/Culture/Representation*, ed. James S. Duncan and David Ley. London: Routledge.

Sommers, Brian J. 2008. *The Geography of Wine: How Landscapes, Cultures, Terroir, and the Weather Make a Good Drop*. New York: Plume.

Street, Richard Steven. 2004. *Beasts of the Field: A Narrative History of California Farmworkers, 1769–1913*. Stanford, CA: Stanford University Press.

Sullivan, C. L. 1994. *Napa Wine: A History from Mission Days to Present*. San Francisco: Wine Appreciation Guild, Ltd.

Treasury Wine Estates. 2012. "About TWE." www.tweglobal.com/about/twe (accessed January 6, 2012).

Tuan, Yi-Fu. 1990. *Topophilia: A Study of Environmental Perception, Attitudes, and Values*. New York: Columbia University Press.

Tyrrell, Ian R. 1999. *True Gardens of the Gods: Californian-Australian Environmental Reform, 1860–1930*. Berkeley: University of California Press.

Ulin, Robert C. 1996. *Vintages and Traditions: An Ethnohistory of Southwest French Wine Cooperatives*. Washington DC: Smithsonian Institution Press.

Unwin, Tim. 1996. *Wine and the Vine: An Historical Geography of Viticulture and the Wine Trade*. New York: Routledge.

———. 2011. "Terroir: At the Heart of Geography." In *Viticulture: The Geography of Wine,* ed. Percy H. Dougherty. Dordrecht, Netherlands: Springer.

Walsh, Carey. 2000. *The Fruit of the Vine: Viticulture in Ancient Israel*. Winona Lake, IN: Eisenbrauns.

Wilkins, John M., and Shaun Hill. 2006. *Food in the Ancient World*. Malden, MA: Blackwell.

Wine Institute. 2012. "The 2011 California and U.S. Wine Sales." www.wineinstitute.org/resources/statistics/article639 (accessed November 14, 2012; page now discontinued).

Winkler, A. J. 1974. *General Viticulture*. Berkeley: University of California Press.

Winter, Mick. 2001. "Who Owns Napa Valley's Vineyards?" In *Wine Business Monthly* (May). www.winebusiness.com/wbm/?go=getArticle&dataId=8217 (accessed December 11, 2014).

Zoecklein, Bruce W. 1995. *Wine Analysis and Production*. New York: Chapman & Hall.

AMITA SINHA and RAJAT KANT

City of Nawabs to City of Elephants

URBAN TRANSFORMATION IN LUCKNOW, INDIA

In the state elections in Uttar Pradesh, India, held in early 2012, the election commissioner ruled that statues of elephants, symbol of the ruling Bahajun Samaj Party (BSP) and Chief Minister Mayawati, be draped so as to avoid unduly influencing the voters.[1] This unprecedented ruling speaks to the power of images in swaying the masses, not surprising given the dominance and impact of figural imagery in the visual culture of India. Statuary has been a very significant element in the large-scale urban transformation of Lucknow, capital of India's most populous state, Uttar Pradesh. Known popularly as the "city of Nawabs," an allusion to its nineteenth-century Muslim rulers who were great patrons of architecture and performing arts,

the distinct cultural identity of Lucknow has survived both colonial urban development and the large-scale expansion of the post-independence period in the latter half of the twentieth century.

Historic Lucknow was oriented to the River Gomti with monumental palaces, mosques, mausoleums, and gardens built on its southern bank between 1781 and 1856 C.E. In the aftermath of an uprising in 1857, bringing colonial rule in its wake, the urban center of Lucknow shifted south with the building of a cantonment, or permanent military station, in proximity to the railway station and civil lines for the colonial bureaucracy. In the postcolonial period since 1947, as the city's population has grown to about three million, the city has expanded in all directions, especially to the north across the River Gomti. Recent large-scale urban insertions in the form of memorial parks, plazas, and streets have created a new look for the city, one that is at odds with its historic Nawabi character and its colonial past.

This chapter interprets the new urban landscape as an expression of the political ideology of BSP that seeks to fabricate heritage for the historically disenfranchised Dalit community. The term "dalit" refers to the untouchable castes that have been socially and economically marginalized for a millennium in Indian society. They constitute 22 percent of Uttar Pradesh's population and, along with other castes low in social hierarchy and termed as "backward," are represented by the BSP, which provides them with a voice and platform. Through their substantial vote bank, BSP is seeking to overturn centuries of exploitation and repression by the higher castes.

A critical reading of this new landscape, an important part of Lucknow's public realm, requires an explicit articulation of values that guides its production, existence, and continuing use. Among the many collective values that are expressed and promoted through the physical environment are those that signify the building of symbolic, social, and environmental capital—three significant dimensions of the public good.

The display of Mayawati's autocratic power in the public sphere provides symbolic capital for the Dalit community, and thus a landscape of empowerment. This landscape of empowerment will be explored through analysis of architectural forms and spaces of memorials.

Social capital reinforces the imagined Dalit community; as settings of everyday social interaction, festivals, and rallies, parks are important catalysts in building face-to-face communities. Social values attached to place-based community and recreational cultures of play, pause, and movement are interpreted from results of a survey of memorial park visitors.[2] Parks, as urban settings where individual or collective encounters with ordered nature occur, should ideally build environmental capital for the city. To examine whether that capital has been achieved in the memorial parks built by Mayawati in Lucknow, this study analyzes specific design features of memorial parks and their reception and assessment by visitors.

Perhaps no other leader has so consistently and vociferously promoted her legacy through such an aggressive program of statue and monument building than Mayawati, who demonstrated a determined personal and political agenda and its ruthless implementation in a surprisingly short time. First elected to power in 1995, she

Fig. 10.1. Satellite image of Lucknow, showing locations of Mayawati's memorial parks. Image annotated by author.

immediately began her building campaign and pursued it relentlessly every time she came into power (1997, 2002, and 2007). Like her other administrative measures, her impressive building record has been controversial, often making headlines in the media and causing public outrage. Because of her ambitious ventures, her detractors have compared her to Mohammad bin Tughluq, the mad sultan of Delhi in the fourteenth century.

A populist by temperament, Mayawati has not hesitated to use every ploy available to promote her party's ideological rhetoric.

Projecting herself and BSP as instruments of social change in the caste-ridden structure of Indian polity, she has pushed for rural development through low-income housing, accessibility to higher education through universities and colleges, and schemes for women in education. The success of these programs is yet to be evaluated, but what is manifestly visible is her effort to inject into the urban space of Lucknow a memorial culture celebrating the successes of social reformers and her own party in improving the condition of Dalits (Sinha 2009).

The memorials—Kanshi Ram Smarak Sthal and Green (Eco) Garden, Buddha Vihar Shanti Upvan, Bhimrao Ambedkar Samajik Parivartan Sthal, Ramabai Ambedkar Maidan, and Prerna Kendra—are all dedicated to BSP leadership. Ambedkar Sthal and Prerna Kendra were begun during Mayawati's third term, the other three in her fourth term (fig. 10.1). With the exception of Prerna Kendra, the other four, designed by Jay Kaktikar of Design Associates, a Noida firm, are located at a distance from the city core on the road from the airport and on the banks of the River Gomti where large empty expanses of land were easy to acquire.

The memorials comprise large precincts of buildings, plazas, parks, streetscapes, and assortments of statuary, fountains, gateways, public conveniences, and parking lots. Their function varies from image gallery (Kanshi Ram Smarak, Prerna Kendra, and Ambedkar Sthal), museum (Ambedkar Sthal), library and monastery (Buddha Vihar), to rally grounds (Ramabai Ambedkar Maidan). Built of expensive materials such as marble, sandstone, and granite, the structures are expected to last for a "thousand years" on "land banks" on the urban fringe, precluding any other urban development. They are largely hardscape urban insertions on a gigantic scale and extend into the busy crossroads of the city. With a clever manipulation of symbols, Mayawati, in keeping with her persona, has succeeded in making defiant monumental gestures that have irrevocably altered the path of the city's growth pattern.

Making History

> I am not inventing history; I am only highlighting history that has been consciously suppressed.
>
> —Mayawati 1997

What kind of history is Mayawati highlighting? Certainly not history as conventionally understood—as representation of the realistic past, objective analysis of empirical data or facts, and occurring in a space and time distinct from that of the historian with the requisite distance that is implied in such an exercise (Hays 2009). She and her architect, Jay Kakitar, are making history through design by building new urban foci. Mayawati appears to be extending social memory of the Dalits through enlarging the "specious present," that is, by bringing the past within the immediate perception (Becker 1932). She and the BSP Party do this through an aggressive promotion of statue-building throughout Uttar Pradesh, thereby enhancing the cult of the *mahapurush* (great soul), whether they are charismatic medieval saints, hero-warriors, or social reformers of the last two centuries. This mythologizing of history—medieval saints and social reformers embarking on the quest for social equality and spiritual enlightenment—resonates with the archetypal myth structure of romance (White 1978).

Scholars have predicted that the memorials and statutes will build an "imagined Dalit community" transcending religious, caste, and linguistic differences. The arrival of Dalit presence in public spaces in the city from where they had been excluded for centuries is empowering, an "incentive for democratic mobilization," a "remarkable tool for the pedagogy of the oppressed," and a "focal point for ceremonies that build grass roots mobilization skills" (Jaoul 2006). Narayan (2011) describes the statues of Dalit local heroes, saints, social reformers, Ambedkar, and Buddha as creating a new visual and oral sphere of memories that together with commemorative rituals are a cultural resource for arousing political consciousness among Dalits.[3] By reclaiming public spaces through Dalit symbols and iconography, Mayawati is not only asserting Dalit identity but also reconstructing collective memory, instilling pride in their past, and helping them gain self-respect. However, this appropriation of the public realm that in theory is a collective space

owned by all, not just the Dalits, represents a segmented, perhaps even a fractured Indian polity.

The reconstruction of cultural memory by bringing the past into the specious present requires not the typical historian's craft, a dispassionate analysis of past events and historic personalities, but the creation of a social milieu within which the glories of the *mahapurush* can be sung, and the metaphorical return of the hero can be reenacted collectively. In this, the built environment becomes the catalyst for facilitating the specious present, providing the vivid richness of the "here and now" in settings for iconography that seeks to instruct and inspire through mimesis. The ensemble of Dalit statuary is part of the monumental complex of buildings, their forecourts, and the urban streetscape. Through its sense of permanence and sheer physical size, the monumental complex celebrates state power and fulfills its original function of reminding one of the exalted role of the Dalit leader in society (Dynes 1996). The making of subaltern memory (not history) of a shared glorious past and its representation in elite urban spaces is a very visible act of empowering the community that had been rendered voiceless and made invisible through erasure from public spaces of urban and rural India for centuries. Memorial parks as sites of memory—*lieux de mémoire*—

reinforce the Dalit collective self, that is, the capacity to imagine, construct, and inhabit a monumental space (Nora 1989; Schane 1996).

Landscape of Empowerment

What exactly is Dalit memory, and how is it inscribed in urban spaces and encoded in architectural forms? Heritage or what is valued from the past is, in a sense, manufactured, residing in symbolic value created through the built environment. This is a form of empowerment of the hitherto powerless Dalit community. Although the urban landscape is only a medium for communicating power and ideology and is not inherently powerful or powerless, it can feel coercive in the experience of its monumentality and the grandeur of its scale. Memorial parks built by Mayawati in Lucknow can be decoded as an empowering landscape/landscape of power in terms of the following features: their location, visibility, enfilade, architecture, scale, statuary, and spectacular affect.

LOCATION AND VISIBILITY

With the exception of Prerna Kendra, all other parks are located away from the urban core and are on the outskirts of the growing city. This ensures availability of vast acreage, prominent locations along the major urban arteries—Buddha Vihar and Kanshi Ram Smarak on VIP/Airport Road, leading to Cantonment and Civil lines, Ramabai Ambedkar Maidan on Ring Road that encircles the city and connects with the state highways, and Ambedkar Sthal on the banks of the River Gomti on the southeast margin of expanding Lucknow. The large cone of visibility from the major traffic arteries, ensured by the absence of high-rise development around the parks, means that the building domes dominate the skyline and command attention over long distances. Their dominant verticality denotes symbolic power, as does the long horizontal stretches of Buddhist-style railings enclosing vast precincts where no other land use is allowed. Their location, size, and imageability give them a landmark status within the city.

ENFILADE

Enfilade, or the linear structure of space as in a sequential arrangement of spatial segments, offers a high level of control over movement and social interaction. Power as discerned in the degree of accessibility is supported by enfilade of rooms in buildings and in nested urban precincts (Dovey 1999). Hillier's (1996) analysis of space syntax is useful in understanding the sequence of urban spaces that provide points of

Fig. 10.2. Moat-like effect of outer precinct; enfilade used in design of the gateway of Ambedkar Sthal. Photograph by author.

potential control over access and plazas at the crossroads. Historic buildings and urban spaces in Lucknow (as elsewhere in the Indian subcontinent) controlled physical access, punctuated movement, and framed vistas through elaborate gateways. The linear series of walled courts in the historic *Imambaras* were accessible though gateways that were thresholds to increasingly private and controlled sacred spaces.

Enfilade is employed deliberately in forecourts in Ambedkar Sthal (fig. 10.2),

Prerna Kendra, Buddha Vihar, and Ramabai Ambedkar Maidan. A series of gateways control access to the inner court of Ambedkar Udyan, Kanshi Ram Smarak Sthal, and Buddha Vihar, entry to which is only possible by purchasing a ticket. The lack of pedestrian access to memorial precincts from streets with heavy, fast-moving traffic creates a moat-like effect upon crossing a series of enclosures which have to be penetrated before the memorial building can be entered. The citadel look is a common device in the layout of twentieth-century national capitals such as in parliamentary complexes designed by Geoffrey Bawa in Colombo, Sri Lanka, and the Capitol Complex by Louis Kahn in Dacca, Bangladesh (Vale 1992).

Mayawati's memorials extend into the main crossroads (*chauraha*) of the city and symbolically appropriate them through statuary in giant plazas. Parivartan Chowk, north of Kaiserbagh and east of Begum Hazrat Mahal Park, an urban insertion by Mayawati in 1995, received negative publicity since it did not fit in the Kaiserbagh Heritage District. A tall vertical structure holding a globe at its summit and sheltering a seated statue of the Buddha marks the center of a huge plaza that acts as a traffic roundabout in the heart of Lucknow (fig. 10.3). Statues of three nineteenth-century social reformers and one of Ambedkar with inscriptions on pedestals mark the cardinal points in the circular plaza where a number of streets converge. This location not only ensures high visibility and landmark status but also marks the arrival of sociopolitical change (new BSP leadership) in a historic area significant to the Nawabi and colonial times.

Monumental plazas such as Bhimrao Ambedkar Chauraha announce the entry to the gateway on the main road from the airport. Located at the convergence of five streets, Samtamulak Chauraha is an ensemble of statues of BSP icons, a symbolic

Fig. 10.3. Monumental elements of Parivartan Chowk, 1995. Photograph by author.

threshold to the kilometer-long spine (part of it a bridge over the River Gomti) punctuated by imposing handsome gateways. These highly visible statue plazas afford panoramic vistas into the urban landscape of converging streets, access to which can be potentially controlled. More significantly, they assert the symbolic presence and gaze of BSP leaders at major nodes of the city.

Fig. 10.4. Buddhist architectural elements in Ambedkar Sthal. Photograph by author.

ARCHITECTURE

With their highly architectural properties, Mayawati's memorial buildings in Lucknow appear to be unique in postcolonial India, where the memorialization of national leaders has largely occurred using the medium of landscape design, not architecture (Sharma 2004). At Rajghat in New Delhi, for example, groves, lakes, and gardens commemorate Gandhi and other Indian prime ministers.

Indeed, the closest parallel to Mayawati's program in its monumental grandeur and iconography seems to be Victoria Memorial in colonial India that created a past for the British Raj (Metcalf 1989).[4]

Rahul Mehrotra (2011) includes Ambedkar Sthal in the category of counter-modernism and resurfacing of the ancient in Indian architecture since the 1990s. Ambedkar's rejection of Hinduism's exploitative and hierarchical caste structure and

his conversion to Buddhism are the *raison d'être* for the Buddhist architecture revival in BSP buildings. The neo-Buddhist style is very evident in Sanchi stupa-inspired domes, boundary walls as Buddhist railings, *chaitya* window relief pattern on walls, freestanding square pavilions, and Ashokan pillar (fig. 10.4).

The adoption of the Buddhist mantle comes via New Delhi where Sir Edward Lutyens (1869–1944) and Herbert Baker (1862–1946) used a neoclassical imperial style of architecture to legitimize the soon-to-be-waning British Empire. The dome of Rashtrapati Bhavan designed by Lutyens, the double-height facade of Baker's Secretariat buildings, along with upturned

saucers, fountains, and flat sheets of water in tanks have been unambiguously copied by Mayawati in three memorial parks— Ambedkar Udyan, Kanshi Ram Smarak Sthal, and Buddha Vihar (fig. 10.5).

While the architecture of imperial Delhi houses government institutions and is literally the seat of legislative and executive power, the sole function of memorial architecture in postcolonial Lucknow is revival/ rebuilding of collective memory. Buddhist and imperial architectural elements imbue the city's memorial buildings with borrowed associations of sacred and political power from India's ancient and recent history, thereby legitimizing the empowerment of the Dalit community in the public sphere.

SCALE

The extra-human scale is created spatially and formally through building structures and elements. At large sites—Ambedkar Sthal (107 acres), Kanshi Ram Smarak Sthal (86 acres), Eco Garden (112 acres), Ramabai Ambedkar Maidan (50 acres), Buddha Vihar (32.5 acres)—are built huge rectangular and long linear plazas, oval and square greens, and a radiating amphitheater. In the vast spatial expanses, there is little to establish human scale and no enclosure except that afforded by a few buildings, separated by large distances. The proximate senses—tactile, olfactory, and haptic—are not stimulated, and vision is the dominant sense in experiencing the physical environment. The eye travels far along visual axes established by linear elements—columns and rows of stone elephants. The long vistas to buildings and other focal points are impressive in their command of physical space and the sense of power that is communicated in gazing at this landscape.

The extra-human scale is employed in buildings as well. The domes soar—for example, the height of Kanshi Ram Smarak building dome is 177 feet and is said to be one of the world's largest. The building plinth is high, making one climb up to the soaring interiors (fig. 10.6). The tall bronze fountains (30 to 52 feet), 18-foot-high marble and bronze statues, high gateways and boundary

OPPOSITE:
Fig. 10.6. Extra-human scale of the plaza; stone elephants line the approach to Kanshi Ram Smarak. Photograph by author.

Fig. 10.7. Buddha statue on Gomti Boulevard. Photograph by author.

walls, massive elephants, the 71-foot-high *stambh* (column) in Ambedkar Sthal, the larger-than-life animals in Eco Garden, all dwarf the individual and reinforce the scale of the buildings. With the exception of palms, there are few trees to bring down the scale or give enclosure. The building textures and the paving patterns in their large size do not relate to the human body. The consistent use of extra-human scale in space and building elements results in diminishing the sense of physical self and making the physical environment appear dominant and powerful.

STATUARY

Mayawati's memorials have been described as an "architecture of statues" (Mehrotra 2011). While buildings frame statues in open spaces within the memorial parks and outside at the city crossroads, statues in plazas and lawns become nodes and termini, creating a system of visual and physical axes that structure urban space and become focal points of vision and movement. The ensemble of four statues facing the four cardinal directions— of Buddha, Ambedkar, Kanshi Ram, and Mayawati—gestures to the concept of *chakravartin*, the world-ruler, whose power radiates out to the horizon. The figures gaze out into the public in an urban *mise-en-scène* that lends a theatrical touch to the urban spaces (fig. 10.7).

The function of these statues goes beyond embellishment of architecture and landscape; they are meant for *darshan* (ritual sighting of the divine in deity, person, or place) and of felicitation rituals on anniversaries and other occasions. Statues of medieval saint poets—Kabirdas, Ravidas, Ghasidas—and of warriors—Birsa Munda—are very visible reminders of greatness achieved in the face of extreme adversity. This deification of great people is deeply rooted in Indic culture and, by converging with the colonial tradition of erecting statues in parks and urban squares, extends the sacred embodied in *mahapurush* iconography into civic spaces.[5]

Mayawati, however, is celebrating not only social reformers and leaders of her party but also her own personality cult. She has justified the erection of her statue in her lifetime (this is not the norm) by claiming that her mentor Kanshi Ram wanted her to. Etchings with vignettes of major events in her life (and that of Kanshi Ram and Ambedkar) are found in memorial interiors, supplementing the freestanding statues in a rich visual archive. This demonstrates a shrewd grasp of how visual culture can be manipulated to garner support, win allegiance and thus votes in a populist democracy, but also implies self-glorification.

With the rise of BSP and Mayawati to power in Uttar Pradesh, statue installations of Ambedkar and other social reformers not only received an impetus, but made folk art into a monument. Little statues of Ambedkar, about fifteen thousand of which were installed in the last two decades in villages, towns, and cities across northern India, have given way to an ensemble of marble and bronze statues in memorials and urban crossroads, signaling the transformation of a community-based movement to a state-sponsored building enterprise with seemingly endless funds at its disposal (Jaoul 2006). The insertion of statuary and its associated structures into elite urban spaces in the city marks their appropriation and expresses the BSP's central agenda of representing symbolic capital in the imagined Dalit community. Other political parties in Uttar Pradesh— Samajwati Party and Lok Dal—have also built parks in Lucknow and dedicated them to their leaders by erecting statues—but not with such abandon as Mayawati, who seemed to have lost all restraint in her obsessive bid to fill the landscape with statuary.

SPECTACLE

Tasteful lighting of buildings, parks, and plazas in memorial precincts creates spectacular scenes at nighttime, enhancing their status as major landmarks and nodes of the city. The multihued lighting of domes, walls, and columns enhances architectural form; that of fountains, columns, and statuary modulates space, recreating the monumental feel when daylight fades. In Gomti Park, the dancing colorful fountains synchronized with music create a dazzling dynamic landscape. Viewers are mesmerized by the show, speaking admiringly of the spectacular effect and taking photographs with their cell phones. The mood created is festive, different from that experienced in daytime, and a major attraction to tourists who choose to come back to the memorial precincts at nighttime. The object of tourists' gaze is a spectacular landscape focused on the glorification of BSP leadership. This landscape communicates the power of Dalit icons and that of the state through colorful dynamic imagery and illusionistic effects. The audience is awed and mesmerized into silence or reduced to social communication focused upon the object of their gaze (Debord 1995). Tourism serves Mayawati's dictum that the parks further her administration's agenda of "Sarvajan Hitay va Sarvajan Sukhay"—benefit and happiness to all (as opposed to "Bahajun," literally meaning "many" but referring to other backward castes)—a manner of "doublethink" in which the pursuit of power is disguised as public service (Dovey 1999).

The rally grounds of Ramabai Ambedkar Maidan remain closed to the public, the only one among the memorial sites so far not open to casual visitors. Its stated function has been to host political rallies only, not to accommodate social events such as marriages and concerts. It unambiguously expresses power enacted periodically in political speeches to mass gatherings of up to 700–800,000 that affirm faith in party ideology and reinforce solidarity among party cadres. Although named "Maidan," denoting a vernacular landscape of large empty grounds serving as public commons, the rally grounds are designed as a paved amphitheater with gently rising tiers fitted within a trapezoidal site (fig. 10.8).

All sight lines focus on the person of the leader addressing the rally from a domed pavilion on a raised platform. Statues of Bhimrao Ambedkar and his wife, Ramabai, under domed *chattris* on circular stepped plazas in trapezoidal lawns flank the central

pavilion. So far only the BSP has organized rallies here, the last one in December 2011, in which Mayawati described her sacrifices for the cause of Dalit uplift, including her decision to remain single. The emotional bonding between the leader and the masses is facilitated by the design and validated by the iconic presence of Ambedkar (and his wife), the author of the Indian Constitution who was responsible for giving the power of the vote to the Dalits.

Symbolic Value

Mayawati's legacy to Lucknow has been an extensive system of plazas and parks that commemorate Dalit identity but also bring the city's residents and tourists in large numbers to gawk at the statuary and memorials. Although she has attempted to encourage archival research and the discipline of history by building museums in Prerna Kendra and Ambedkar Sthal and a library and lodging for Buddhist monks in Buddha Vihar, the fabrication of Dalit heritage was Mayawati's most important

agenda. She has done it primarily through iconographic representations of Dalit saints, social reformers, and leaders in the public space—building symbolic capital embodied in the personal charisma of the *mahapurush*. This may appear as an anachronistic gesture given that memorial parks in postcolonial India have followed an abstract and modern design vocabulary, but it is in keeping with the age-old Indic ethos of celebration through figural imagery (Sharma 2004).

The landscape vocabulary of the memorial parks/plazas is one of power established through extra-human scale, neoclassical and revivalist architecture, citadel effect, and spectacle. In this Mayawati and her architect, Jay Kaktikar, have fallen back on a universal idiom of power encoded in the built environment by autocratic, imperial, and even democratic regimes. Imperial Delhi of the British Raj, Chancellery, the proposed Reichstag in Berlin and Nazi rallies in Nuremberg, and other national capitals such as Washington D.C., Canberra, and Islamabad, to name a few, have employed some or all of the above elements in a symbolic display of power and legitimation of

Fig. 10.8. Rally grounds of Ramabai Ambedkar Maidan. Photograph by author.

authority. While these are or were functioning centers of administration, the memorial landscapes of Lucknow are heritage sites of the newly constructed subaltern past. Their symbolic value overrides their social and environmental values.

Like other imperial rulers in India's history—the great Mughals and the British—Mayawati has sought to leave a legacy in stone. She has used the same means as the elite and privileged who sought to leave a mark through extensive building programs and indeed has done a remarkable job. This fabricated heritage was embodied in neo-revivalist Buddhist architecture of ancient India, not Lucknow's historic hybrid Indo-Islamic and colonial architecture, although it did employ the nested urban precinct idea from its urban past. Nawabi Lucknow of Imambaras and the old city exists in an uneasy juxtaposition with the Sanchi domes, elephant galleries, and Buddhist ornamentation and railings. While this layering reveals the passage of time, it nevertheless interferes with the image of a visually coherent city.

Needless to say, the media have over-whelmingly reacted negatively to the memorial parks, calling the statues "just one clue as to the extraordinary cult of personality that has grown up around the 'Dalit Queen.'"[6] Columnists and bloggers have called attention to the vast amount of public funds that were spent on questionable projects: "the extravagancy of Ambedkar Sthal à la Taj Mahal appears a mockery of people living in funds-deprived Bundelkhand and Poorvanchal regions of the State."[7] Others have called her the "Lutyens of Lucknow," asserting that "her giant parks must rank among the greatest new public spaces created in any Indian city since independence" (Swamy 2012).

A survey of fifty randomly chosen visitors conducted in August 2012 showed interesting results: it is seen as a place to learn Dalit history, but only half of the respondents at Ambedkar Sthal agreed that it represents their heritage; only one-fifth believed that to be true of Kanshi Ram Smarak and that it raised the profile of Dalits; although Ambedkar Sthal was rated highly as a place to learn Dalit history, so was Buddha Vihar. Eco Garden and the riverfront boulevard received the lowest rating in this regard, with a majority saying that they did not contribute to Dalit heritage.

One of the disliked features of Ambedkar Sthal and Buddha Vihar was Mayawati's statue elevating her to the level of Lord Buddha. The parks, however, were overwhelmingly seen as improving Lucknow's image through their architecture, plazas, gardens, and streetscapes as well as their connection with the inserted monumental *chaurahas*. The resistance to acknowledging memorial parks as symbols of Dalit heritage is curious, given the very visible presence of Ambedkar and BSP statuary inside memorial halls and in plazas and streets. Since the majority of respondents were non-Dalits, perhaps the explanation lies in the lingering antipathy and the refusal by higher castes to give Dalits their due in the public sphere. The symbolic value of parks appears to lie primarily in enhancing Lucknow's image, not in memorializing Dalit heritage, thus only partially fulfilling Mayawati's intentions.

Social Value

The public realm fosters both imagined and place-based communities and helps build social capital. Parks are as much about communion with nature as they are about spontaneous and planned interactions with other people (Cranz and Boland 2004). The social value of parks is an important dimension of the public realm that underlies the sense of place identity. Mayawati's goal was to increase tourism through spectacular effects and Buddhist-revival architecture in memorial parks. Most respondents to the survey agreed that the parks did indeed bring tourists. Do the parks facilitate recreational subcultures of exercise, play, and festivals that together with informal socializing are responsible for building local communities?

The spaces are designed for viewing buildings, statues, fountains, and other objects, and it is not surprising that photography is a popular activity. The range of activities is not large since the parks lack features that would promote play and interactions with the environment. People engage mostly in walking, sitting, and talking with friends and family. Young couples seeking privacy in public spaces dominate. Lack of public transportation to the parks and the fact that they are large sites surrounded by streets with heavy vehicular traffic preclude daily visits for morning and evening walks and informal socializing that make small neighborhood parks such people-friendly places (Marcus and Francis, 1990; Sinha 2005). Lack of food courts and vendors also means that social gregariousness revolving around eating together is absent. Respondents in all parks wanted seating, play facilities, drinking fountains, and food vendors. These features can be installed to increase the social value of the public parks since they not only fulfill immediate needs but also promote sociability. Both indoor and outdoor spaces can be programmed for marriage and other celebrations. For example, the Meditation Hall in Buddha Vihar, already a popular place with the elderly, can host lectures and sermons that draw the public in large numbers.

Fig. 10.9. Green (Eco) Garden. Photograph by author.

Environmental Value

The new memorial landscape is not environmentally friendly. The parks do not have much tree cover, and the large built-up mass creates a heat-island effect requiring large bodies of water in hot summers to cool the microclimate. They are popularly known as "stone gardens" because of their vast expanses of paving, presumably built to avoid the maintenance costs incurred in keeping the nature in parks under control. Instead of lawns, a nonsustainable resource-consumptive landscape, other kinds of ground cover could have been used in Kanshi Ram Smarak, Eco Garden, and Buddha Vihar. The Gomti Parks, built on the floodplain, neither physically connect with the river nor address its ecology. Eco Garden appears to be designed as a Jurassic theme park with exotic flora, larger-than-life fauna, and greenhouses (fig. 10.9).

Visitors to the Ambedkar Sthal, Kanshi Ram Smarak, and Buddha Vihar are conscious of the absence of environmental values in the design of the parks. All three were rated low in being environmental friendly and as places to enjoy nature. When asked what they disliked about the parks, visitors remarked about lack of canopy trees, excessive use of marble, and lack of

greenery. Eco Garden, however, was rated highly on environment friendliness but did not score as well on its natural value. The riverfront boulevard between Ambedkar Sthal and the Gomti was also rated low on its environmental friendliness and presence of greenery. Visitors want better views of Gomti and more trees on the boulevard.

Landscape management of Gomti Parks and Eco Garden will require huge resources that the city may not choose to commit. Besides environmental sustainability, social sustainability is also questionable as the parks are too large to be appropriated and maintained by locals. Environmental values did not guide memorial park design; however, they could be added, keeping in mind the expressed needs of park users for shade and connection with the river. Tree planting in all parks will increase greenery and canopy trees, while providing shade will also reduce the scale and create a sense of smaller spaces that can be easily appropriated. Some of the best-used public spaces can be found on the Gomti floodplain acting as *maidans*; the lawns over time can become sites for cricket and other games and public gatherings. Although it would be difficult to restore floodplain ecology, it may yet be possible to build *ghats* (steps) and piers that would bring people down to the river and actively engage with it (Nagpal and Sinha, 2009).

Conclusion

Mayawati and the BSP were voted out of power on March 6, 2012. Within a week, a new government was formed by the Samajwadi Party, with the support of OBC (other backward castes) and Muslims, 21 percent and 29 percent of Uttar Pradesh's population, respectively. Akhilesh Yadav, the current chief minister, had contemplated building medical colleges, research and IT centers in the BSP memorial parks. But the reality is that parks cannot be altered or building structures demolished readily due to the extensive use of stone and building technologies such as concrete-raft foundations in a predominantly hardscape environment. Mayawati and her architect have ensured that dismantling will occur at a huge cost and will likely incur the wrath of Dalits, who from all accounts take inordinate pride in the monuments.

In building this landscape of empowerment, Mayawati was accused of abusing her power as the chief executive of the state. Diversion of state funds, pressing the state institutions into service, and giving building contracts to Dalits and OBC have been criticized, but none are in themselves illegal acts. Resistance through filing by public interest litigation (PIL) on the part of citizens resulted in Supreme Court rulings in her favor. Critics say that Mayawati's government showed democracy at its worst because it brought an autocrat into power, although voting her out of power in the 2012 election has demonstrated that democracy in India works within the context of caste politics, the driving factor being balance of power among all caste groups in the long run. It could be argued that there are other and more effective ways of empowering the subaltern through education, government employment, and slum-rehabilitation schemes. Given that 85 percent of India's population falls in the OBC category, the needs are vast and the resources never enough to fulfill the demand.

The changed political reality requires a critical examination of the complex set of interactions among symbolic, social, and environmental values and how they may be negotiated in planning and design of public spaces. Finding a center is ensuring convergence of multiple values for promoting the public good in the ongoing, dynamic process of urban landscape change. The recent spate of memorial parks in Lucknow in the past decade memorializing regional parties and their leaders has highlighted the segmented nature of its public realm (Sinha 2010). Few lessons for creating cohesive public landscape can be found in going back further in time. The Nawabi gardens were private enclaves accessible to the wealthy and privileged; the colonial park was also exclusive to the European and Indian gentry (Sinha 1996; Sinha and Kant 2000). Parks in the postindependence era did seek to memorialize national leaders and place identity, but those earlier efforts have given way to contested and conflict-ridden urban landscapes, particularly along the Gomti riverfront towards the southeast. Mayawati attributed her party's defeat to Muslims voting for Samajwadi Party; perhaps if she had erected memorials to Nawabi Lucknow she may not have lost so heavily. As it is, her central mission of memorializing Dalit leaders through development of public spaces designed around statuary in the state capital has been successful and will likely endure. Adding social and environmental value to the memorial parks, hitherto focused exclusively on buttressing Dalit symbolic capital, will make them more inclusive and be a step towards repairing the deep fissures in society.

NOTES

The authors are grateful to the late R. P. Sinha and J. P. Sinha for their help in the fieldwork, to students in Gautam Buddh Technical University, Lucknow, for assisting in the survey, and to David Hays (University of Illinois at Urbana-Champaign) for sharing readings from his Making History seminar.

1. A teacher by profession, Mayawati joined politics in 1984 after meeting Kanshi Ram, founder of the BSP, representing Dalits (oppressed people belonging to the untouchable castes) and other lower backward castes. The party's strongest base is in Uttar Pradesh where *chamars* (leatherworkers) form a substantial minority of OBC (Other Backward Castes) and to which Mayawati belongs. She rose swiftly through the party ranks and was elected its leader within a decade. Her rapid ascent to power speaks to her political acumen, strong determination, and excellent leadership skills.

2. Visitors in Ambedkar Sthal, Kanshi Ram Smarak Sthal, Gomtifront Boulevard (popularly known as Marine Drive), Eco Garden, and Buddha Vihar were surveyed in August 2012 by students in Government College of Architecture, Lucknow. Fifty visitors, randomly selected, responded to questions about symbolic, social, and environmental values, and improvements they would like to see in the memorial park.

3. Bhimrao Ambedkar is the author of the Indian Constitution that gives equal rights to all citizens, irrespective of caste or religious affiliation. Ambedkar belonged to a low caste by birth and converted to Buddhism in protest against Hinduism's caste hierarchy.

4. Victoria Memorial in Calcutta, the capital of colonial India, was designed by Sir William Emerson in the Indo-Saracenic architectural style—a combination of Palladian and Indo-Islamic forms—in the early 1900s. It commemorated Victoria, empress of India, in exhibition galleries arranged around a central hall, capped by a soaring white marble dome.

5. The lifelike representations lend the famous name a vivid image, in keeping with the dual significance of name (*naam*) and image (*roop*) in Indic religious thought and political discourse (Davis 1997). Iconographic representations provide a visual object for veneration and ritual commemoration.

6. *Hindustan Times,* March 14, 2012.

7. *Economic Times,* March 6, 2008.

REFERENCES

Becker, Carl. 1932. "Everyman His Own Historian." *American Historical Review* 37 (January): 221–36.

Cranz, Galen, and Michael Boland. 2004. "Defining the Sustainable Park: A Fifth Model for Urban Parks." *Landscape Journal* 23 (Fall): 102–20.

Davis, Richard. 1997. *Lives of Indian Images.* Princeton, NJ: Princeton University Press.

Debord, Guy. 1995. *The Society of the Spectacle.* Trans. Donald Nicholson-Smith. New York: Zone Books.

Dovey, Kim. 1999. *Framing Places: Mediating Power in Built Form.* London: Routledge.

Dynes, Wayne. 1996. "Monument: The Word." In *"Remove Not the Ancient Landmark": Public Monuments and Moral Values,* ed. Donald Martin Reynolds, 27–31. Amsterdam: Gordon and Beach.

Hays, David. 2009. "Making History." In *Envisioning the Bloomingdale: 5 Concepts (Chicago Architectural Club Journal),* ed. Clare Lyster, 111–12. Chicago: University of Chicago Press.

Hillier, Bill. 1996. *Space Is the Machine: A Configurational Theory of Architecture.* Cambridge, UK: Cambridge University Press.

Jaoul, Nicholas. 2006. "Learning the Use of Symbolic Means: Dalits, Ambedkar Statues and the State in Uttar Pradesh." *Contributions to Indian Sociology* 40 (2): 175–207.

Marcus, Clare Cooper, and Carolyn Francis, eds. 1990. *People Places: Design Guidelines for Urban Open Space.* New York: Van Nostrand Reinhold.

Mayawati. 1997. Interview by *India Today* (August 11), 33.

Mehrotra, Rahul. 2011. *Architecture in India since 1990.* Mumbai, India: Pictor.

Metcalf, Thomas. 1989. *An Imperial Vision: Indian Architecture and Britain's Raj.* Berkeley: University of California Press.

Nagpal, Swati, and Amita Sinha. 2009. "The Gomti Riverfront in Lucknow, India: Revitalization of a Cultural Heritage Landscape." *Journal of Urban Design* (UK), vol. 14 (November): 489–506.

Narayan, Badri. 2011. *The Making of the Dalit Public in North India: Uttar Pradesh, 1950–Present.* New Delhi: Oxford University Press.

Nora, Pierre. 1989. "Between Memory and History: Les Lieux de Mémoire." In *Memory and Counter-Memory*, special issue, *Representations* 26 (Spring): 7–24.

Schane, Murray. 1996. "The Psychology of Monuments." In *"Remove Not the Ancient Landmark": Public Monuments and Moral Values*, ed. Donald Martin Reynolds, 47–52. Amsterdam: Gordon and Beach.

Sharma, Yuthika. 2004. "Memorial Design in India." *Journal of Landscape Architecture* (India), vol. 9 (Spring): 17–21.

Sinha, Amita. 1996. "Decadence, Mourning and Revolution—Facets of the 19th-Century Landscape of Lucknow, India." *Landscape Research* 21 (2): 123–36.

———. 2005. "Reinventing People's Place." *Architecture + Design* (India), vol. 22 (September): 54–58.

———. 2009. "Public Spaces in Lucknow: The Influence of Power." *Architecture + Design* (India), vol. 26 (February): 90–92.

———. 2010. "Colonial and Post-Colonial Memorial Parks in Lucknow, India: Shifting Ideologies and Changing Aesthetics." *Journal of Landscape Architecture* (Europe), vol. 10 (Autumn): 60–71.

Sinha, Amita, and Rajat Kant. 2000. "Urban Evolution and Transformations: Study of Medieval and Colonial Streets in Lucknow." *Architecture + Design* (India), vol. 17 (September–October): 66–71.

Swamy, Subramaniam. 2012. "Behenji's Raj: Mayawati, the Lutyens of Lucknow." *Times of India* (February 26).

Vale, Lawrence. 1992. *Architecture, Power, and National Identity.* New Haven, CT: Yale University Press.

White, Hayden. 1978. "The Historical Text as Literary Artifact." In *Tropics of Discourse: Essays in Cultural Criticism*, 81–100. Baltimore: Johns Hopkins University Press.

DON MITCHELL and TOM MELS

From Gotland to Youngstown

THE INDISSOLUBLE LINK BETWEEN LANDSCAPE AND JUSTICE

Gotland

The landscapes of Gotland, off the east coast of Sweden in the Baltic Sea, are as remarkable as they are varied. It would be hard to call them spectacular—the morphology of the island does not really allow for that—but they are impressive. From the prehistoric huts (fig. 11.1), compounds, and Tings that dot the interior of the island to the medieval-cum-Hanseatic-cum-postindustrial city of Visby, from the old beet-sugar processing plants (that provided quick energy for the working classes of industrializing Europe) to the massive limestone quarries that provide the cement for much of the rapid transformation of cities that marked Europe's

pre-2008 boom (fig. 11.2), from the converted wastelands now turned into funky tourist sites to the rural retreats of Princess Eugenie in the nineteenth century and Håkan Nesser in the twenty-first, and from the endless fields of forage and wheat to the mucky mirelands from which they were sometimes carved (fig. 11.3), an amazingly varied scene unfolds over a relatively small space.

To understand the Gotland landscapes as they exist now, however, it is important to understand precisely how capital penetrated the seemingly peripheral countryside of Gotland, and the ways in which this process roused an acute awareness among residents of how struggles over landscape and its transformation are always struggles over

Fig. 11.1. Prehistoric hut preserved on the island of Gotland. Photo by D. Mitchell.

Fig. 11.2. One of Europe's largest limestone quarries, northeastern Gotland. Photo by D. Mitchell.

Fig. 11.3. Reclaimed mireland, now planted in rapeseed, central Gotland. Photo by D. Mitchell.

justice. As Tom Mels (a Gotlander) has shown, British entrepreneurs in the middle nineteenth century worked closely with the Swedish state—itself interested in the clearance of mirelands and their conversion into agricultural landscapes—to redistribute mireland from the commons they had long been to privately held, exchangeable tracts of *property* better suited to what we would now call clear-cutting. Capital was poured into the land, radically transforming its physical geography; it was also poured into the hands of newly proletarianized workers, radically transforming the social relations through which they lived. The advent of state-sponsored, British-capital-led land reclamation induced a move, as David Harvey has put it more generally, from a "world in which 'community' is defined in terms of structures of interpersonal social relations to a world where the community of money prevails" (Harvey 2010, 294). This was a fundamental, foundational switch. Never divorced from the world of trade and capital—the importance of Visby in the Hanseatic League[1] gives lie to that—nonetheless, up until the middle of the nineteenth century, most villages and peasants lived more communal lives based in mutual labor, the commons, and use-value, than would ever be possible afterwards.

Transformations of such depth never pass unchallenged. After all, mirelands

Fig. 11.4. Industrial
Youngstown in its
heyday, ca. 1919.
www.brackenridge.net.

were places of ingrained use-value, shaped through customary fishing and hunting rights, haymaking, and a whole array of mutual obligations. Alienation from family and common land through deceit and violence led to significant struggles to reclaim the means of production—and through the courts, their land; workers rebelled. Coupled with these struggles, the evident environmental destruction that marked the reclamation of mirelands, the clear-cutting, and the transformation of hay meadows into industrial-scale beet fields, meant that the remaking of Gotland into a specifically *capitalist* landscape was a fraught, lengthy, and in fact always incomplete process (Palmenfelt 1994, 126).

Youngstown

The landscapes of Youngstown, Ohio, along the banks of the Mahoning River could not be more different from those of Gotland. They are as spectacular as they are typical.

It would be hard to call them remarkable: they are little different from any number of other deindustrialized cities in the Rust Belt. They are little different, but they are iconic, and now they have become iconic for the third time. From the end of the nineteenth century through the first three-quarters of the twentieth, Youngstown was a quintessential steel town. Massive mills stretched for twenty-four miles up and down the Mahoning and coal smoke filled the air (fig. 11.4); tight-knit ethnic neighborhoods climbed the hills away from the river, even as land was set aside to create large landscaped parks (Linkon and Russo 2002). Home to Youngstown Sheet and Tube, Youngstown was one of the high-tech cities of its time. While class struggles always simmered just below the surface and sometimes broke out into the open, and while race and ethnic divisions were keenly deepened and exploited by mill owners and managers, smoke meant prosperity (Bruno 1999). Youngstown was an icon of mid-twentieth-century industrial might.

By the early 1980s, Youngstown had been totally deindustrialized. Youngstown Sheet and Tube (now owned by the New Orleans–based conglomerate LTV) led the slaughter in 1977; other companies quickly followed suit (Buss and Redburn 1983). The effect on working-class people in the town was, of course, devastating (Camp 1995; Lynd 1982).

Fig. 11.5. Shrinking Youngstown. Source of photo unknown.

Distress was everywhere apparent in the landscape (fig. 11.5): houses were abandoned, arson soared, stores closed, bars as well as schools were rapidly shuttered, the gorgeous landscaped parks went unmaintained. Youngstown became an icon of the ravages of footloose capital—indeed of the new mobility and thus the new violence that finance capital was able to wreak. The absence of smoke might mean a cleaner environment; but it also meant devastation—near utter devastation (High and Lewis 2007). The withdrawal of capital in Youngstown was a wrenching transformation every bit as significant and as violent as the entrance of capital into Gotland more than a century earlier. Youngstown's landscapes were an icon of this moment of capital's undevelopment exactly as it was an icon of earlier moments of capital's industrial might.

Now Youngstown has become iconic again, its landscape symbolic for a third time.

It is at the forefront of a global "shrinking cities" movement. City officials are seeking to shrink its footprint, rationally planning a smaller city, wherein some neighborhoods are abandoned altogether and others are more tightly focused around key institutions (such as a school, a church, or a small shopping center). Efforts to attract new capital are focused almost entirely on downtown and the area near the university. With these efforts, Youngstown has become a model, its redevelopment strategies the focus of intense study and increasing emulation (Hassett 2008). Youngstown is iconic, now, of what many take to be a revolutionary move in America: planning not for capital's local growth, but for its ongoing retrenchment. Yet the city's success in attracting reinvestment downtown while seeking to clear out other districts has come at a cost of a decreasing standard of living for most Youngstowners. There are now more renters than homeowners in what used to be called the "City of Homes," and absentee landlordism is a problem. Crime and unemployment rates remain astronomically high. The notion of a rationally shrinking city, in other words, seems little more than smoke and mirrors, as the remnants of the working class seem to have little to gain from the remaking of Youngstown's landscape (Russo and Rhodes, n.d.).

These two examples—Gotland and Youngstown—confirm in different ways that justice and injustice are embedded in, maintained, and contested through the landscape. Land reclamation on Gotland— the reclamation that marked Gotland's entry into modern capitalism—was a process of primitive accumulation, "the historical process of divorcing the producer from the means of production," as Marx put it (1954, 668). For Marx, primitive accumulation was a process of robbery, of theft—robbery and theft that are the necessary but unjust foundation for the free exchange of goods in capitalism. As a case of primitive accumulation, the transformation of mire landscapes implicates justice in a broad spatial sense, ranging from the obliteration of customary practice, to efforts by the dispossessed to reclaim their right to land, and on to experiences of environmental damage.

In the more urban setting of Youngstown, the long history of investment, disinvestment, and partial, highly localized reinvestment— of capital circulating through the built environment—has led to different levels of working-class power, and thus to vastly uneven outcomes. Many of those thrown out of work during Youngstown's deindustrialization found themselves stuck. The capital sunk in *their* landscapes—their homes and churches and parks—was essentially permanent; unlike the capital held by Youngstown Sheet and Tube's parent company, LTV, it could not get up and go. It could only be devalued in place, destroyed as a precondition for new capital investment—and that on a much smaller, and highly uneven, footprint (Harvey 1982). The landscape that has resulted is not necessarily any more, or less, just than that which preceded it, but that landscape *has* resulted in a rearrangement of relations of power on the ground and transformed the conditions of social struggle.

Why Justice?

Why talk about landscape in the language of justice? The landscapes of Gotland and Youngstown indicate that the link between landscape and justice, while complex, is also indissoluble. Justice is central to exactly what landscape *is*. Justice is an ideal; landscape, however, is always more than an ideal. It is the very real material effect— the "spoor" as Peirce Lewis (1975) called it—of social practice, as well as the basis for that practice. It is the built form of the world "as it really is" (Harvey 2001), even as, through representation, the meanings of that world are highly varied and deeply powerful (W. J. T. Mitchell 1994; Duncan and Duncan 1988). Landscape is heavy, solid, powerful, concrete. Justice, by contrast, seems as immaterial as it is aspirational. As Engels wrote: "Justice which is the organic, regulating, sovereign basic principle of societies, which has nonetheless been nothing up to the present, but which ought to be everything—what is that if not the stick with which to measure all human affairs, if not the final arbiter to be appealed to in all conflicts?" (quoted in Merrifield and Swyngedouw 1995, 1).

All three iconic forms of Youngstown's landscape can easily be measured against this stick: at each moment of its evolution, the landscape there called up and made solid in its morphology questions of equity—the just distribution of the spoils of capitalist development and the burdens (often bodily) of capitalist destruction (N. Smith 1990). Similarly with the mirelands: use-values were embedded in the landscape, and their alienation, expropriation, and destruction both required and advanced a reorientation of those use-values, that is, who would benefit from this reorientation and who would lose.

It required remaking the landscape and thus the very nature of the space within which "justice" (of whatever sort) could and could not be achieved. In remaking the landscape, a new notion of "the good" was instantiated in and through the landscape, which those behind the expropriation of the commons and its remaking into private property understood to be refracted through notions of progress, economic development, and profit making— a very different sense of what constituted "good" and "right" than had operated earlier among the peasants of Gotland.

But what is this "stick" to which Engels referred? It can have a specifically liberal form. John Stuart Mill asserted in 1863: "Justice is the name for certain classes of moral rules, which concern the essentials of human well-being more nearly, and are therefore of more absolute obligation, than other rules for the guidance of life" (quoted in D. Smith 1994, 23). This merely specifies the question: what are these "classes of moral rules," and how are they decided? David Smith argues that "justice involves treating people right or fairly, in a calculated way," and identifies two different forms: *retributive justice*, which concerns the fair imposition of penalties on those who commit crimes or sin against prevailing norms; and *distributive justice*, which entails the fair allocation of goods (1994, 24). Distributive justice is usually further divided into the overlapping categories of *economic justice* (the fair distribution of costs and benefits), *social justice* (the fair distribution of life-chances), and *political justice* (the fair distribution of power, rights, and liberties). Justice—Mill's "classes of moral rules"—can also be categorized as *procedural justice* (fair rules) and *substantive justice* (fair outcomes). These two, of course, can be—and frequently are—contradictory; procedurally just processes can lead to substantively unjust outcomes, just as substantively just outcomes can be created through procedurally unjust processes (Fainstein 2010).

Mainstream liberal discussions of justice are typically confined to questions of distribution and procedure (D. Smith 1994). John Rawls's (1971) famous thought experiment—wherein people find themselves at the dawn of society and must divide up available goods without knowing who will get to choose their share first and who last, and who thus parse out benefits as close to equally as possible—is perhaps the classic statement of distributional justice. Against Rawls, however, the libertarian philosopher Robert Nozick (1974) notes that goods to be shared out come from somewhere; they are produced. Nozick thus "considers 'just' a distribution in which individuals keep whatever good they can accumulate" through their own production and trade (Mansbridge 1995, 363).

Rawls and Nozick present variants of liberal philosophies of justice, and as such they incorporate particular notions of procedural justice. Rawls begins from a presumption that all members of society are free and equal, and thus are able to enter contractual arrangements as equal agents; power does not intrude. A procedurally just society is one in which such freedom and such absence of power is manifest. The role of the state in such a world is to guarantee an "even playing field" upon which autonomous, equal, free individuals may meet. For Nozick, whose ideas were rooted in a variant of Lockeanism, any self-owning individual ought to be free to use her own, or to appropriate unused resources, just so long as no other individual was made worse by such use or appropriation (Vallentyne 2010). A procedurally just society, in this view, is one in which individuals are at liberty to use, or to not use, what is rightfully theirs in any way the individual pleases. The only role for a state in such a world would be to guarantee the conditions of liberty. There is, in this view, little room for the sort of customary forms of justice that structured the mirelands as commons, much less for the dispossessed workers of Youngstown Sheet and Tube who sought in 1979 to seize the property of LTV—the means of production—and put them to their own use as LTV off-shored its capital.

It is not hard, then, to see the limits of

either liberal version of justice. More radical theorists of justice therefore find them wanting, not only for their inattentiveness to power, but also to their ignorance of the very real social relations of class, race, and gender that always shape distribution. There *is no* "original position," like Rawls imagined; there *is no* radical individualism like Nozick posited—both of which lead to an uncritical acceptance of prevailing institutional contexts and exploitative relations of production. Against such liberal positions, therefore, Iris Marion Young argues that "social justice" (understood as something more than only the just distribution of life chances, though this is a necessary condition) is the most encompassing form of justice (1990, 15). She understands social justice to mean "the elimination of institutional domination and oppression." Young is not as concerned with individuals as she is with social groups, which "are not entities that exist apart from individuals, but neither are they arbitrary classifications of individuals according to attributes which are external to or accidental to their identities " (44). That is to say, neither Rawlsian contractualism nor Nozickian libertarianism is sufficient for understanding substantive possibilities for, and procedures of, social justice, especially since "groups . . . constitute individuals," rather than vice-versa: individuals are not atomistic and ontologically prior to the groups they are part

of. Custom—and solidarity—matter. So does power.

Social justice must therefore be theorized in relation to the group (and the individuals in it), and for Young (1990) it must be theorized in relation to oppression—and, as I will suggest, in relation to the landscape and what it makes possible or impossible. Rawls's "original position" makes no sense because the world is always already here, and always already uneven, unjust. The libertarianism of Nozick makes no sense because, as Susan Okin (1989) has put it, "all possession is socially constructed. Women [as a group] produce children through hard labor, but do not conclude they have property rights in their product" (Mansbridge 1995, 364). Women, of course, labor in a lot of other ways, frequently to find the products of that labor expropriated by others; gendered divisions of labor (as just one example) mean already unequal access to power, and hence some in society, as groups, find themselves in *structurally* oppressive circumstances. As Young trenchantly put it: "The freedom, power, status, and self-realization of men is possible precisely because women work for them" (1990, 50). Oppression, in other words, is the other side, the hard side, of Engel's "stick."

As Young famously argued, oppression, the hard side of the stick, has five faces—not just exploitation of the type just described, but also marginalization, powerlessness, cultural imperialism, and violence. For Young, marginalization is different from exploitation in that it begins not from appropriating the fruits of the labor of others, but by excluding others from the very ability to productively labor at all. What Marx called primitive accumulation and what David Harvey (2003) has reformulated as "accumulation by dispossession" relies on such marginalization. To the degree that mire-people were uprooted from their land, their labor made redundant, then to that degree they were marginalized from the rising system of production, at least until such time as their very presence as a laboring class (rather than a class rooted in customary access to the resources of the landscape) had been assured.

Exploitation and marginalization are made possible by the development of powerlessness. But powerlessness exceeds the other two, especially in modern society, since "the labor of more people [not just the traditional working class] augments the power of the relatively few," as Young put it (1990, 56). And powerlessness is manifest in other ways as well. For example, some factions of capital in Youngstown—the local capitals who had the bulk of their wealth "trapped" or "fixed" in the landscape of small shops or stores, for example—simply could not pick up and move as could the capital encapsulated in the plants and productive processes of LTV. "The powerless," according to Young, "are situated so that they must take orders and rarely have the right to give them"; they are subject to, rather than masters of, the forces that shape them.

Cultural imperialism is oppression of a different form. It "involves the universalization of a dominant group's experience and culture, and its establishment as the norm" (Young 1990, 59). In Nancy Fraser's (1987) view (but Young's [1990, 59] words), such oppression means that some groups are subject to, rather than the controllers of, "the means of interpretation and communication in society" such that groups differing from the norm find their experiences and ways of knowing "reconstructed largely as deviance and inferiority."

Finally, oppression may manifest itself as violence. Young points in this regard to "random, unprovoked attacks on person or property," but the invocation of randomness, as her subsequent discussion in fact makes clear, is misleading: oppressive forms of violence are *systemic*. They might be "low-level," such as "incidents of harassment, intimidation, or ridicule [conducted] simply for the purpose of degrading, humiliating, or stigmatizing group members," or they might be concerted campaigns against targeted groups, such as lynchings, gay-bashing, or mass rape. In either case, the violent acts are systemic and made possible by "the social context surrounding them, which makes them possible and even acceptable" (Young 1990, 61).

To this sort of overt violence must be added another sort of systemic violence, a form that is especially important in relation to the question of landscape and values (Loyd 2009; Mitchell 2010): what anthropologists call "structural violence." Structural violence, in Paul Farmer's words, is "suffering [that] is 'structured' by historically given (and often economically driven) processes and forces that conspire—whether through routine, ritual, or, as is more commonly the case, the hard surfaces of life—to constrain agency" (2003, 40; Galtung 1969), leading to premature and excess death, high rates of disablement, chronic illness, and the like.

Landscapes are the hard surfaces of life—the very *place* of injustice or, indeed, of justice. Invoking Marx, David Harvey (1996) argues for a theory of justice that understands that prevailing and conventional standards of social formations—including especially their geographies—condition what is right and what is just. "Right can never be higher than the economic structure of society and its cultural development conditioned thereby," Marx and Engels argued (1970, 19). And in this view, as a *historical proposition,* the surplus value-producing, waged-labor relation at the very heart of capitalism—

wherein a worker gives up for a set time the right to the use of her labor power *and that which it produces* in exchange for an agreed-upon wage—is perfectly just, even if, with Young and Marx, we might still call this "exploitation." As Marx wrote: "The justice of transactions between agents of production consists in the fact that these transactions arise from the relations of production. . . . The legal forms in which these economic transactions appear as voluntary actions of the participants, as the expressions of their common will and as contracts that can be enforced on the parties . . . by the power of the state, are mere forms that cannot themselves determine this content. They simply express it. The content is just so long as it corresponds to the mode of production and is adequate to it. It is unjust as soon as it contradicts it. Slavery, on the basis of the capitalist mode of production, is unjust; so is cheating on the quality of commodities" (1981, 460–61).

But within capitalism, the exploitation of wage-labor is totally just. What is just—*justice*—is rooted in specific histories, specific geographies; it is not universal, not transhistorical. Yet Marx nonetheless manages to condemn capitalism as unjust, invoking a normative critique of capitalist property rights, of the theft by which they were made possible, of the appropriation of surplus value, and of exploitation itself,

Fig. 11.6. Windmills in southern Gotland. Although cooperatively owned and governed, they give no hint as to the relations of production, ownership, use, and power that govern their presence. Photograph by D. Mitchell.

sounding in the process more like Iris Marion Young or Nancy Fraser than the Marx famously dismissive of bourgeois values. David Harvey follows suit: any sense of the historicity of justice, he argues, must be tempered with arguments about what justice *could be* under other historical-geographic conditions. To achieve a *different kind* of justice, he implies (2009, 137), one would not attend (for example) merely to the question of "exploitation"—to what it is and why it is bad as measured against some external stick (an idealist effort)—but instead to the conditions under which *particular forms* of exploitation are advanced, and how they can be transformed so that some other way of living and producing can be achieved (a normative, historical-materialist effort).

The Politics of Landscape

The landscape is precisely *where* multiple forms of oppression—and thus opportunity—coalesce (Harvey 1996, 349). The material shapes and structures of landscape, however, are not simply already-given exterior surfaces. They are the contested results of human labor, human labor that under capitalism is fully exploited labor. Yet landscapes are frequently subject to what Rich Schein calls "an amnesia of genesis" (Schein 1997, 663). Rather than yielding themselves up,

the historical geographies of justice and injustice through which landscapes come to embody that labor—that power—may be hidden behind a physical appearance of naturalness or seemingly neutral depiction and description: landscape is ideology in built form.

Like space generally, the immediate perspicuity of landscape "is illusory and the secret of the illusion lies in the transparency itself" (Lefebvre 1991, 287). This is why critical geographers seek to unpack reified notions of normality, inevitable naturalness, or rationality through which landscapes appear as devoid of human labor, power relations, and political process, in order to trace the specific uneven conditions under which landscapes are materially and discursively produced (Mels 2002; D. Mitchell 1996). By extension—because this necessarily foregrounds the discursive and material workings of oppression and dominance—such a project means that landscape is unavoidably caught up in notions of justice and injustice (fig. 11.6).

But landscape is not *only* ideology in built form. The struggle over primitive accumulation on Gotland offers a useful entry point to think about this. The geographer Kenneth Olwig has argued that there is a central, historical struggle between place-based logics and practices of landscape on the one hand, and geometric, cartographic space-based logics and practices on the other (2002a). Each implies a different political orientation towards landscape (as well as different landscape forms)—and thus different notions of political justice: the former, place-based practices, are based in "custom centered on the particularity of place"; the latter, cartographically shaped practices, are based in universal laws, or "nature de-centered in universal space" (2005, 299–300).

In Europe, place-based practices of landscape, Olwig shows, were rooted in "the sort of enclosed room-like area that is demarcated, for example, by the territories of historically constituted places such as the German *Landschaft* polities," which were more-or-less autonomous, self-governing *common* polities that stood against or perhaps apart from the more rigid feudal polities that were dominant (2002b, 3). They were rooted in custom and the sort of deliberation marked by the presence of a "thing" (ting)—a parliament or deliberative body (fig. 11.7). Gotland was a *landskap* in this sense, with principles of justice codified in its own medieval customary laws but under constant threat of the centralized imperial ambitions of the Swedish and Danish states. Much later, in the wake of primitive accumulation, the abstract space of land-reclamation maps, scientific-

knowledge production, and elite power ran into such place-oriented notions of justice and conventional practice, and it was this collision that shaped much of the landscape we now see.

Landscape is a form of representation but, as Olwig shows, historically it has been a *place* of *political* representation, a site where natural and customary law is both instantiated and practiced. For Olwig, then, landscape-as-place is a *preferred* site of justice. As he writes, "place is primary in terms of historical and personal experience. . . . It is the moral behavior that arises through the sharing of common resources which has the greatest effect in building communities because it is rooted in the habitus of practice" (2005, 318). Hence Olwig's recurring emphasis on the contrast between place, custom, and convention on the one hand and the rationality of cartographic space on the other hand: "The meaning of landscape, in practice," Olwig avers, "is . . . very much a regional affair, with roots going deep into the identities of . . . historical regions. . . . This is the landscape of place in which people become attached to history" (2009, 203).

For people in Gotland and Youngstown, however, landscape has always been *more* than a regional affair. People in these places became "attached to history" not just through

Fig. 11.7. Remnants of a thing (or ting), northeastern Gotland. Photograph by D. Mitchell.

some secluded, rooted, and meaningful landscape of place, but rather through a restless network of extended social relations and spatial processes that produced those places as part of the uneven landscape of capitalism. Any understanding of "customary" landscape, then, must be aware of how custom—always a form of justice—is implicated in and developed out of spatial experiences of (for example) imperialism, anticolonialism, and the dynamics of a global capitalist system (Mels 2006).

In this sense it is worth thinking for a moment about one of the most important developments in the politics of landscape, and thus in the evolving relationship between landscape and justice, in recent times: the creation of the European Landscape

Convention. This convention reveals a penchant for understanding landscape as a common resource, a place of community and customary rights, albeit in awkward tension with outspoken top-down tactics that threaten to ensconce landscape in an exclusionary version of justice. The former tendency is rooted in the sort of progressive politics of difference that Iris Marion Young advocates, while the latter turns that politics into the sort of conservative defense of the status quo that David Harvey (1996) worries about.

Approved by the Council of Europe in 2000 and coming into effect in 2004, the European Landscape Convention promulgates the recognition in law of landscape "as an area, as perceived by people whose character is the result of action and interaction of natural and/or human factors." The convention obliges "Contracting Parties"—that is, signatory European states—"to recognize landscapes in law as an essential component of people's surroundings" and identities; "establish and implement policies aimed at landscape protection, management, and planning"; create procedures for public participation in management and preservation; and "integrate landscape into regional and town planning policies" as well as into policies covering other practices (agriculture, industrial development, and so forth) that might affect the landscape

(Déjeant-Pons 2006, 370). Of course, it also puts questions of design front and center.

The mandate for public participation is both particularly important in relation to normative visions of a just landscape and particularly tricky. Despite its recognition of local culture and lay knowledge, the convention simultaneously supports an approach to procedural justice that prioritizes instrumental rationality through expert rule over the landscape. In the terms of the convention, public participation in landscape policy and management "should not be seen as a substitute for official decision making but as a complement to it. The objective is to draw into the decision-making process the views of all concerned groups of stakeholders, whether defined as local communities, residents, visitors, landholders, deprived groups, or specialists, alongside representative, democratically elected bodies" (Jones 2009, 234; cf. Olwig 2009).

As attractive as that sounds, Michael Jones (2009, 237–38) has noted that "the Convention's own Explanatory Report recommends 'performing the evaluation [of the landscape] according to objective criteria first' (as if any criteria can be objective), and then comparing the findings with the assessments of the landscape by people concerned and other interest groups." The problem is that "expert discourses usually dominate and effectively exclude"—

marginalize—those less than fluent in it (Aitken 2002; cf. Fainstein 2010). They can be a form of cultural imperialism—to use Young's term—that through their avowedly expert-led scientism serve to mask fundamental, substantive concerns since "conflicts concerning landscape values are often symptomatic of deeper-lying social conflicts" (Jones 2009, 248; Jones 1999).

With its technocratic faith in expertise, combined with a celebration of place-identity and difference, the European Landscape Convention—important as an effort to protect and preserve difference within a globalizing, capitalist Europe—nonetheless ends up instantiating a specifically neoliberal mode of justice. This is perhaps most clear in the way it selectively appropriates "community" and turns it into a form of "public-private partnership" and form of governance in which *polity* is decidedly absent (Peck and Tickell 2002, 390). Indeed, the selective appropriation of the idea and fact of community in the convention, combined with its paradoxical treatment of procedural justice, seems to leave little room for considerations of the forms of oppression that tend to structure "justice" in capitalist landscapes. As large a step forward as the European Landscape Convention represents, it is not without its infirmities for developing a just landscape, justly arrived at (cf. Harvey 2009, chap. 3).

Conclusion

Indeed, something like the European Landscape Convention seems unimaginable in a place like Youngstown, and imagining the American landscape, shot through as it is with so many forms of social, economic, and environmental injustice, as a series of regional *res publica* (as Olwig's recent work urges) seems even more remote. And yet, as George Henderson insists, imagining such things—and struggling to put them in place—is precisely what is needed: "What we need is a concept of landscape that helps point the way to those interventions that can bring about much greater social justice. And what landscape study needs even more is a concept of landscape that will assist the development of the very idea of social justice. To achieve this geographers and other landscape analysts will need to engage in a more sustained conversation with the disciplines of moral and political philosophy concerning the enumeration of basic human rights and modes of their defense" (2003, 196).

This essay is an attempt to open such a conversation among landscape architects and their allies. In that effort, we join good company—for example, the conversation launched in 2008 in Cambridge (United Kingdom) at a conference on "The Right to Landscape." Bringing together an impressive array of legal scholars, landscape architects, building architects, geographers, and human rights practitioners, the workshop debated the relationship between human rights and the landscape, and specifically sought "to collectively define the concept of the right to landscape and to generate a body of knowledge that will support human rights" (CCLP 2010; quoted in Egoz et al. 2011, 2). This initiative has the potential to galvanize debate around the question of landscape, justice, and social activism in the same manner that the growing enthusiasm around the concept of "the right to the city" has.

While "The Right to Landscape" workshop—and subsequent book—was quite cognizant of the representational aspects of landscape (and how landscape is itself a space of representation, as Ken Olwig insists), it is apparent that any "concept of landscape that will assist in the development of the very idea of social justice" must be more than cognizant of it. The politics of representation— Fraser's and Young's politics of "cultural recognition"—must be worked right into studies *and practices* of landscape if justice is to be served. Rich Schein (2006, 2009, 2012) has made the urgency of such a project apparent in his recent studies of the erasure and reclamation of alternative histories of racialized landscape in and around Lexington, Kentucky. As he makes clear, understanding and bringing to light both the representational and the morphological

erasures in the landscape help to address the workings of cultural imperialism—the flip side of cultural recognition. This is an issue of aesthetics—most certainly—as much as it is one of physical space. As George Henderson argues: "any definition of the beautiful landscape would have to include the full participation of all and the economic means to do so" (2003, 197; Duncan and Duncan 2001, 2004).

Another way to put this is to say that landscape *is* the stick—it is the hard surface of life. As the citizens of Gotland and Youngstown know only too well, landscapes are material and representational evidence— in the rebuilt quarries and abandoned mills, in the reforested mirelands and the struggling neighborhoods, in the historic downtowns and the brand-new redevelopments, in the common polity of the *landskap* as well as the alienated polity of a landscape nearly totally controlled by the whims of footloose capital— of what the current state of justice is, *and* it is the foundation for what justice can become.

NOTES

1. This refers to the role of the city as a trading center in the Hanseatic League—an alliance of Baltic cities and city-states that ruled trade in the North Atlantic and Baltic regions from the thirteenth to the seventeenth centuries.

REFERENCES

Aitken, Stuart C. 2002. "Public Participation, Technological Discourses and the Scale of GIS." In *Community Participation and Geographic Information Systems,* ed. William J. Craig, Trevor M. Harris, and Daniel Weiner, 357–66. London: Taylor and Francis.

Bruno, Robert. 1999. *Steelworker Alley: How Class Works in Youngstown.* Ithaca: ILR Press.

Buss, Terry F., and F. Stevens Redburn. 1983. *Shutdown at Youngstown: Public Policy for Mass Unemployment.* Albany: State University of New York Press.

Camp, Scott D. 1995. *Worker Response to Plant Closings: Steelworkers in Johnstown and Youngstown.* New York: Garland Publishers.

Déjeant-Pons, Maguelonne. 2006. "The European Landscape Convention." *Landscape Research* 31:363–84.

Duncan, James, and Nancy Duncan. 1988. "(Re) reading the Landscape." *Environment and Planning D: Society and Space* 6:117–26.

——. 2001. "The Aestheticization of the Politics of Landscape Preservation." *Annals of the Association of American Geographers* 91:387–409.

——. 2004. *Landscapes of Privilege: The Politics of the Aesthetic in an American Suburb.* New York: Routledge.

Egoz, Shelley, Jala Makhzoumi, and Gloria Pungetti. 2011. "The Right to Landscape: An Introduction." In *The Right to Landscape: Contesting Landscape and Human Rights,* ed. Shelley Egoz, Jala Makhzoumi, and Gloria Pungetti. London: Ashgate.

Fainstein, Susan. 2010. *The Just City.* Ithaca, NY: Cornell University Press.

Farmer, Paul. 2003. *Pathologies of Power: Health, Human Rights, and the New War on the Poor.* Berkeley: University of California Press.

Fraser, Nancy. 1987. "Social Movements vs. Disciplinary Bureaucracies: The Discourse of Social Needs." CHS Occasional Paper No. 8. Center for Humanistic Studies, University of Minnesota.

Galtung, Johan. 1969. "Violence, Peace, and Peace Research." *Journal of Peace Research* 6: 167–91.

Harvey, David. 1982. *The Limits to Capital.* Chicago: University of Chicago Press.

——. 1996. *Nature, Justice and the Geography of Difference.* Oxford, UK: Blackwell.

——. 2001 [1984]. "On the History and Present Condition of Geography: A Historical-Materialist Manifesto." In *Spaces of Capital,* 108–20. New York: Routledge.

——. 2003. *The New Imperialism.* Oxford, UK: Oxford University Press.

——. 2009 [1973]. *Social Justice and the City.* Athens: University of Georgia Press.

——. 2010. *The Enigma of Capital and the Crises of Capitalism.* London: Profile Books.

Hassett, Wendy L. 2008. "The 'Shrinking' Strategy of Youngstown, Ohio." In *Building the Local Economy: Cases in Economic Development,* ed. D. Watson and J. Morris. Athens, GA: Carl Vinson Institute of Government.

Henderson, George. 2003. "What (Else) We Talk about When We Talk about Landscape: For a Return to the Social Imagination." In *Everyday America: Cultural Landscape Studies after J. B. Jackson,* ed. Chris Wilson and Paul Groth, 178–98. Berkeley: University of California Press.

High, Steven, and David Lewis. 2007. *Corporate Wasteland: The Landscape and Memory of Deindustrialization.* Ithaca, NY: ILR Press.

Jones, Michael. 1999. "Landskapsverdier som konfliktpunkt i planlegging: Eksempler fra Trondheim." In *Landskapet vi Lever i: Festskrift til Magne Bruun,* ed. M. Eggen, K. Geelmyden, and K. Jørgensen. Oslo: Norsk Arkitekturforlag.

———. 2009. "The European Landscape Convention and the Question of Public Participation." In *Justice, Power and the Political Landscape,* ed. Kenneth R. Olwig and Don Mitchell, 231–51. London: Routledge.

Lefebvre, Henri. 1991. *The Production of Space.* Oxford, UK: Blackwell.

Lewis, Peirce. 1975. "Common Houses, Cultural Spoor." *Landscape* 19 (2): 1–22.

Linkon, Sherry Lee, and John Russo. 2002. *Steeltown U.S.A.: Work and Memory in Youngstown.* Lawrence: University of Kansas Press.

Loyd, Jenna M. 2009. "'A Microscopic Insurgent': Militarization, Health and Critical Geographies of Health." *Annals of the Association of American Geographers* 99:863–73.

Lynd, Staughton. 1982. *The Fight against Shutdowns: Youngstown's Steel Mill Closings.* San Pedro, CA: Singlejack Books.

Mansbridge, Jane. 1995. "Justice." In *A Companion to American Thought,* ed. Richard Wightman Fox and James T. Kloppenberg, 361–65. Oxford, UK: Blackwell.

Marx, Karl. 1954. *Capital: A Critique of Political Economy.* Vol. 1. Moscow: Progress Publishers.

———. 1981. *Capital.* Vol. 3. Harmondsworth, UK: Penguin.

Marx, Karl, and Friedrich Engels. 1970. *Selected Works.* Vol. 3. Moscow: International Publishers.

Massey, Doreen. 1994. *Space, Place, and Gender.* Minneapolis: University of Minnesota Press.

Mels, Tom. 2002. "Nature, Home and Scenery: The 'Official' Spatialities of Swedish National Parks." *Environment and Planning D: Society and Space* 20:35–54.

———. 2006. "The Low Countries' Connection: Landscape and the Struggle over Representation around 1600." *Journal of Historical Geography* 32:712–30.

Mels, Tom, and Don Mitchell. 2013. "Landscape and Justice." In *The Wiley-Blackwell Companion to Cultural Geography,* ed. Nuala Johnson, Richard Schein, and Jamie Winders, 209–24. Malden, MA: Wiley-Blackwell.

Merrifield, Andy, and Erik Swyngedouw. 1995. "Social Justice and the Urban Experience." In *The Urbanization of Injustice,* ed. Andy Merrifield and Erik Swyngedouw, 1–17. London: Lawrence and Wishart.

Mitchell, Don. 1996. *The Lie of the Land: Migrant Workers and the California Landscape.* Minneapolis: University of Minnesota Press.

———. 2010. "Battle/Fields: Braceros, Agribusiness, and the Violent Reproduction of the California Agricultural Landscape during World War II." *Journal of Historical Geography* 36:143–56.

Mitchell, W. J. T., ed. 1994. *Landscape and Power.* Chicago: University of Chicago Press.

Nozick, Robert. 1974. *Anarchy, State, and Utopia.* New York: Basic Books.

Okin, Susan Moller. 1989. *Justice, Gender, and the Family.* New York: Basic Books.

Olwig, Kenneth R. 2002a. *Landscape, Nature, and the Body Politic: From Britain's Renaissance to America's New World.* Madison: University of Wisconsin Press.

———. 2002b. "The Duplicity of Space: Germanic 'Raum' and Swedish 'Rum' in English Language Geographical Discourse." *Geografiska Annaler* 84 (B): 1–17.

———. 2005. "The Landscape of 'Customary Law' Versus That of 'Natural Law.'" *Landscape Research* 30:299–320.

———. 2009. "The Practice of Landscape 'Conventions' and the Just Landscape: The Case of the European Landscape Convention." In *Justice, Power and the Political Landscape,* ed. Kenneth R. Olwig and Don Mitchell, 198–212. London: Routledge.

Palmenfelt, Ulf. 1994. *Per Arvid Säves möten med människor och sägner: Folkloristiska aspekter på ett gotländskt arkivmaterial.* Stockholm: Carlsson.

Peck, Jamie, and Adam Tickell. 2002. "Neoliberalizing Space." *Antipode* 34:380–404.

Rawls, John. 1971. *A Theory of Justice.* Cambridge, MA: Harvard University Press.

Russo, John, and James Rhodes. N.d. "A Renaissance for Whom? Youngstown and Its Neighborhoods." CWCS Publications. Youngstown, OH: Center for Working Class Studies, Youngstown State University. cwcs.ysu.edu/about/news/renaissance (accessed March 13, 2012).

Schein, Richard H. 1997. "The Place of Landscape: A Conceptual Framework for Interpreting An American Scene." *Annals of the Association of American Geographers* 87:660–80.

———. 2006. "Digging in Your Own Backyard." *Archivaria* 61:91–104.

———. 2009. "Belonging through Land/Scape. *Environment and Planning A* 41:811–26.

———. 2012. "Urban Form and Racial Order." *Urban Geography* 33 (7): 942–60.

Smith, David M. 1994. *Geography and Social Justice.* Oxford, UK: Blackwell.

Smith, Neil. 1990. *Uneven Development: Nature, Capital, and the Production of Space.* 2nd ed. Oxford, UK: Blackwell.

Vallentyne, Peter. 2010. "Libertarianism." In *The Stanford Encyclopedia of Philosophy,* plato .stanford.edu/archives/fall2008/entries/ libertarianism/ (accessed November 18, 2014).

Young, Iris Marion. 1990. *Justice and the Politics of Difference.* Princeton, NJ: Princeton University Press.

M. ELEN DEMING

Finding Center

DESIGN AGENCY AND THE POLITICS OF LANDSCAPE

Questions raised in this book are intended to challenge current conventions and showcase new perspectives on landscape and values. These essays feature specific instances where landscape values are being negotiated with fascinating results still very much in play in the social and environmental imagination. The authors offer fresh insights on that ageless, pliable, and never-more-vital topic. Given that springboard, this concluding chapter directly addresses professional landscape actors, shapers, and designers in order to spotlight the special agency they exercise in negotiating the politics of landscape values by and through design. As Don Mitchell and Tom Mels have suggested in their essay, "The politics of

representation—Fraser's and Young's politics of 'cultural recognition'—must be worked right into studies *and practices* of landscape if justice is to be served" (216).

Readers will easily detect my bias; that I am a landscape architect is obvious both in my intellectual orientation and in references to specific authors and organizations. However, the comments that follow are emphatically meant to include designers from many disciplines equally—from architecture, urban design, engineering, planning, historic preservation, ecological restoration, and many forms of agriculture, in addition to landscape architecture—for all the reasons suggested in the preceding essays. And because *all* people are landscape actors and agents at one level or another, everyone is invited to listen in on the conversation and adopt a share in reading and thus renegotiating the environment.

Coda

Before launching into polemic, however, a brief review and synthesis are in order. The introduction to this volume, "Value Added," suggests how these essays might be framed and understood within the various subgenres of cultural landscape studies, as well as in the context of the concept of cultural materialism. By offering a vantage different from that of the original designer or planner, the authors illustrate the potentialities, the complexities, and even the limits of materialist interpretations.

The nuanced "naturalizing" narratives of dominant culture are explained and criticized in Britton's iconological analysis of the class consciousness of Napa Valley vineyards; in Mitchell and Mels's comparison of the history of landscape injustices wrought by concentrated capital in Gotland, Sweden, and Youngstown, Ohio; and in London's essay detailing the negotiation of ideological symbols and visitor experience in the design of the Korean War Veterans Memorial. Critical case studies enable us to notice, recognize, and perhaps give a name to the

mechanisms of dominant cultural values in many places and landscapes. Landscape professionals may deploy similar techniques that reveal and deflect dominant cultural forces in play in personal and political positions.

Presenting temporal extensions of landscapes past into landscapes present, both residual and emergent cultural values are also explored in these essays. Seavitt Nordenson's story of the de-domestication and subsequent re-wilding of the Oostvaardersplassen exposes the complex ironies of residual values—using genetically manipulated animals as proxies in the struggle to maintain cherished historical narratives, political myths, and environmental values. Certainly, Sears's cultural history of the Midwest shows the genesis and maintenance of residual values in the ways that working landscapes have shaped the normative culture of this region. And, in the design submissions for the Oklahoma City Memorial competition, the vivid spectrum of cultural values analyzed by Holland range from reactionary/residual to visionary/emergent expressions.

By contrast, Meyer articulates an aesthetic theory that suggests a reconciliation of the affect of built and natural environments in our experiences as landscape travelers and designers. This theory preserves extant strains of phenomenology (see Meyer 2001 and 2008), a highly mutable philosophy in design, and helps us recognize the persistence of cultural landscape values even as we move into new realms of highly contemporary expressive possibilities.

Alternative and oppositional (critical) values are also examined. By revising techniques commonly used in community engagement, Brown seeks to create an alternative pedagogy of planning and design as well as to empower marginalized communities in Los Angeles. Sinha and Kant critique a complex building program of politically motivated urban iconology aimed at empowering the Dalit caste in Lucknow, India—essentially an opposition movement. Moore's polemical argument opposes dominant design pedagogy in favor of alternatives to creative research and design investigation.

Each of these essays peels back multiple layers of values embedded within public institutions and engaged processes and shows how they are reproduced in permanent structures—intellectual structures (such as curricula) or physical structures (such as monuments and cities). The authors also demonstrate how social values—rather than being mere abstractions—are actively encoded and operational in real landscapes, and can reveal our own unwitting complicity in their maintenance. If not directly instrumental, analyses and debates of these case studies are consequential and desirable

ends in themselves. Armed with greater capacity to see how landscape performs through its hidden value systems, citizens thus gain a better grasp on the value-laden puzzles that so many landscapes present.

Landscape Values and Conflict Resolution

Revelatory techniques explored in this book may be useful in resolving social conflicts—both in and through landscape design processes. Since the dawn of time, landscapes have been shaped by the expression of conflicting values: conflicts between species (salmon and bears; tigers and men), disputed territorial boundaries between human tribes, sacred rivers used for waste disposal, and so on. An intractable conflict over land resources could be reframed more simply as one theory of goodness, for instance "global sustainability," in conflict with another, for instance the "sovereign rights of nations." Like politics, landscape is local: most landscape values are contested at the local or even personal scale. Indeed, most people understand a landscape only to the extent that it affects their own experiences, resources, and opportunities. However well intentioned or "good," many small-scale and short-term interests can manifest "badness" at larger national or global scales, and vice-versa.

To resolve landscape conflicts, what needs to be resolved first is the gulf between value systems—those vantage points that Meinig identified long ago in "The Beholding Eye" (1979). This begins with the early recognition and articulation of community values and choices, not after the legal semantics are finally parsed. Rendering problems into component values may enable negotiators to focus on satisfying shared goals, rather than pitting personal or corporate positions against each other. This notion holds just as true for landscape issues as, for example, industry/labor relations.

In the ensuing discussions, cultural landscape values may be messy and uneasily expressed: ideas will shimmer between now and then, here and there, hopes and fears, personal and collective values. But the way we inhabit the environment is an essential conversation that all citizens should have an opportunity to participate in. As these essays show, without first having that conversation, no form of social peace, environmental justice, or sustainability is possible. Sadly, the costs of not solving certain problems— water and food insecurity, pandemics, armed conflicts, sea level rise, and so on—usually increase over time. From the oil crisis of the 1970s to the aftermath of September 11, 2001, to the economic collapse of late 2008, international events have influenced both social capacity and political will to act on

questions of social and environmental values. Yet the most intractable landscape problems seem to be the reciprocal effect of political impasse: in other words, if there is no willingness to identify shared values, it is unlikely that effective solutions will be found.

In principle, American processes of governance (administrative, legislative, and judiciary) are structured to acknowledge and balance multiple value systems (systems of "goodness") for the national environment. Because political processes tend to focus on short-term (one election cycle) and small-scale (local or regional) concerns, however, the potential to provide for the "greater good" is not always realized. "Good" values may be contested and/or rejected for equally "good" reasons. Whether such plans are or are not enacted, they ultimately reflect the negotiation and the costs of conflicting values—the essence of politics—in the shape of the landscape.

Let's consider a familiar debate: should we not protect the greater good of the environment along with all its diverse populations? Must people always come first? And are these two goals contradictory? Some will hear this as absurd: of course the needs of people come first! Others will point out that, when people are responsible for endangering the lives of other creatures that play important roles in preserving the health of the total environment, human activities ought to be reined in. Both positions are stated pragmatically. But because the debate is essentially ideological at its heart, each position is "unthinkable" to the other side.[1] When either side proposes that its views should shape regulatory policy, the opposing side will register political resistance.

Examples of divisive environmental ideologies abound. Let's consider downstream pollution, where nutrient runoff from agricultural zones in the American Midwest is shown to be a powerful contributing factor in contaminated drinking water (Toledo, Ohio, summer 2014), or hypoxia (the "Dead Zone") in the Gulf of Mexico (Nassauer, Santelmann, and Scavia 2007). Natural fluvial processes in the Great Lakes and Mississippi drainage basins provide services to one industry (agriculture) while other livelihoods (water supply, fishing) suffer the consequences. Whose responsibility is it to look after both the ecological *and* economic health of the whole system? Which constituents ought to pay the price for remediation or, alternatively, for the actual value/cost of services rendered by Nature? And where are the elected officials capable of making the tough political and economic decisions to restructure costs and benefits in this situation?

Former vice-president of the United States Al Gore has long used his broad visibility to raise awareness of the consequences of global climate change, first by publishing *Earth in the Balance* (1992) and later writing/inspiring the documentary film/book combination *An Inconvenient Truth* (2006). Gore and many others have emphasized the political and moral repercussions of climate change in coastal zones, which are expected to have disproportionate impacts on poor and geographically vulnerable people as well as fragile or threatened species (UNDP and Malik 2014, 1). Gore assigns responsibility for climate change solutions to all citizen-consumers and urges everyone to make choices that will help slow or change the environmental trajectory. But despite international acclaim for his work (including two Academy Awards), it has not led to widespread political will to action in the United States.[2]

President Obama, too, has made efforts to shape domestic responses to the threat of climate change; his 2013 Climate Action Plan sets out an ambitious agenda, framed as a "moral obligation to future generations," of profound interest to all environmental actors and agents (Executive Office 2013, 4). The plan addresses greenhouse gas emissions and anticipates adaptations for and mitigations of the impacts of global climate change, while it also sets in motion strategies for new investments in energy production, distribution, and transportation technologies. But even though "Congressional action is not required to implement the various aspects of the plan, Congress still must agree to fund it" (Lewis-Burke 2013, 1, 9). We can safely assume the political decision-making process surrounding environmental issues such as these is likely to remain ideological and adversarial.

Professional Agency

The good news is that professional agents and designers have enormous potential to participate more broadly in political discussions about values and landscape change. Why be shy? Most environmental scientists and design professionals are trained and positioned to understand, illustrate, and mitigate the unintended (or simply unexamined) environmental consequences of sociopolitical value systems. Yet without engagement and support from an informed, landscape-literate citizenry, professional efforts alone are often rendered ineffectual.

PUBLIC HEALTH, SAFETY, AND WELFARE

In the United States, advocates for several design professions—groups such as the National Council of Examiners for Engineering and Surveying (ncees.org) and National Council of Architecture Registration Boards (www.ncarb.org)—monitor the administration of professional regulation by individual states. The statutory purpose and standards of these organizations are aligned almost invariably "to protect the health, safety, and welfare of the public" (NCEES 2014). The Council of Landscape Architectural Registration Boards (CLARB) has a more landscape-specific purpose: "to foster the public health, safety and welfare related to the use and protection of the natural and built environment affected by the practice of landscape architecture" (CLARB 2009). Statutory protections are thus mandated for a wide variety of professional project types and ostensibly include protection of landscape values. Unfortunately, challenges to professional licensure (seen as exclusionary by some) often come from competing professional groups as well as legislators seeking to streamline "regulatory burdens" on the free exercise of trade. Without clear definitions for value-laden terms or performance metrics, professional registration boards regularly find themselves challenged to defend professional expertise, measure individual competence, and justify professional licensure—one law and one state at a time.

To address this, in 2010, the Council of Landscape Architectural Registration Boards commissioned a content analysis to discover, describe, and illustrate ways that landscape architecture contributes to *public welfare*—a concept with multiple meanings ranging from economic and political interests to non-material interests, morality, and social order (ERIN 2010, 6). "[T]he challenge," they explain, "is to develop a definition of a concept that is recognized as pivotal but that lacks structure. Like truth or beauty, we intuitively recognize public welfare as a desirable quality, yet have some difficulty pinning it down" (3).[3]

Based on content from standard professional texts and journalism in the field, CLARB's analysis identified seven distinct forms of public welfare—that is to say, measures of value added by design to landscape:

- Enhances environmental sustainability
- Contributes to economic sustainability
- Promotes public health and well-being
- Builds community
- Encourages landscape awareness and stewardship
- Offers aesthetic and creative experiences
- Enables people and communities to function more effectively.

Concluding that the concept of "public welfare is [still] open-ended and developmental," the CLARB study challenges practitioners to "positively impact public welfare and create more beautiful, functional, and sustainable outdoor spaces," as well as to

identify methods that may more effectively measure their impact on public welfare (ERIN 2010, 70, 74–76).

At first glance, these seven categories seem promising; they certainly offer a necessary and useful point of beginning. Comprising a predominantly anthropocentric theory of "goodness," however, these values appear to operate within typical scales and conventional interests in urban (or urbanizing) environments. What happens when we start to examine the terms more critically? Where were textual referents sought?[4] Whose well-being or aesthetic experiences are included? What forms of function or sustainability are protected, and which are not? And so on.

Local community examination of these values might provide a platform for larger discussions and surely ought to lead to the articulation of richer local categories of values that transcend the initial study. But because any meaningful or articulate measures of value will be closely tailored to the intricacies of local value systems, they may also signal the presence of alternative, oppositional, and/or emerging theories of goodness (see discussion of Raymond Williams, this volume, 17–19). Yes, much easier said than done.

CORE VALUE(S) IN DESIGN

Let's take this train of thought deeper into the field of landscape architecture.

The essay "Most Important Questions" (1992), published in *Landscape Journal* by then-editors Robert Riley and Brenda Brown, asked twenty or so contemporaries: "What do you consider the most important question(s) in landscape architecture today?" This remarkable challenge resulted in a collection of responses that Brown and Riley described as "mind-numbing in their range and diversity." Defying their hopes for a clear diagnosis, they found instead that "[a]ny generalizable observations, neat conceptual categories, insightful analyses, or shrewd speculations about where we are and where we are heading proved elusive at best" (160).

The same year, the University of Virginia took up the gauntlet by hosting the annual meeting of the Council of Educators in Landscape Architecture and taking as their theme *Design + Values* (Rosenberg 1993). One of the keystone events at that conference was a panel designed to continue the conversation moderated by the editors of *Landscape Journal*, including landscape architect and educator Reuben Rainey, who raised some provocative questions of his own: "Given that landscape architecture is a design and planning profession, and is the expression in built form of values or policies, what are the ultimate values that inform our professional work? From this question comes a series of subquestions: What relationship do these values have to the great philosophical and religious traditions that have shaped human consciousness throughout history? Can these values be proven empirically or are they statements of faith? What is the relationship of beauty to value, if any?" (Brown, Riley et al. 1993, 38).[5]

In pondering the deeper moral compass that guides professional practice, Rainey also teeters on the brink of challenging the ways such practices are taught and learned. What moral obligation does a designer have to the processes of the natural world and to other species? How far, or how long, is the community willing to stretch ecosystem services and concepts of environmental justice in order to expand and maintain the infrastructure of human existence? What tradeoffs are clients and their professional consultants willing to make, and why? No doubt readers can contribute their own pressing questions and arguments to a growing list. Let it suffice to say that, over the years, these and many other efforts comprise a series of occasional studies examining two essential, reciprocal issues in design and values: (1) understanding how social values guide designers' work, and (2) seeking measurable evidence of the value of that work to society.

How Social Values Guide Professional Work

A few years after the Virginia conference, in direct response to the "Most Important

Questions" quandary, Ian H. Thompson responded with the book *Ecology, Community, and Delight: Sources of Values in Landscape Architecture* (2000). "Far from asking for more technical knowledge," Thompson wrote, "the majority of the published replies [in "Most Important Questions"] were concerned with values" (2). Thompson decided to pursue a more nuanced rendering of the range of values that landscape architects typically express through their work and, in the process, to confront the inevitable pesky multiple constructions and synonyms for the word *value*—for instance the terms "goodness," "worth," "principles," or "ethics" (3–4). Despite his initial hope to identify an overarching central value that might bind all ecological, social, and aesthetic issues together, what Thompson discovered was such "a multitude of divergent values . . . it became apparent that to force them into some artificial structure or rigid hierarchy was going to do violence to some of them" (7).

It didn't prevent a brave attempt, however. Katherine Crewe and Ann Forsyth (2003) classified a variety of approaches evident in a sampling of landscape architects' work as profiled in the pages of *Landscape Architecture Magazine* from 1997 to 2001.[6] The authors analyzed eight facets of each work: project goals, process, client/audience, scale, intellectual base, ethics, approach to nature, and an analysis of power relations.

According to Crewe and Forsyth, *LandsCAPES* (the name of the typology they developed) "provides a framework for analyzing practice as a *value laden process* . . . and the core values that [landscape architects] express" (original emphasis, 37). Their purpose is to help practitioners "reflect upon and debate dimensions of the profession that are too often implicit and invisible. For example, do landscape architects see their work as politically neutral or as a force advocating the preservation of the natural world? Does landscape practice promote sustainability . . . ? Do particular approaches promote social equity or maintain social hierarchies?" and so on (37–39). Awareness of these approaches, say the authors, "potentially makes explicit six quite different paths to excellence in the field . . . [and] points to areas where education can be reformulated to value different paths in practice" (51).[7]

Ostensibly still seeking to identify those values and expertise that legitimize the discipline and its basis for intellectual autonomy, *Harvard Design Magazine* recently begged a new version of the old question in a theme issue entitled *Landscape Architecture's Core?* (Summer 2013). But the very notion of "core" is challenged by Ian Thompson. In "Essence-less Landscape Architecture and Its Extended Family," Thompson characterizes landscape architecture as "a broad church" in which there are fundamentalists and agnostics, puritans and skeptics, essentialists and nonessentialists, each identifying their issues and, in the process, continuously redefining what is central to the field. As Thompson puts it: "As the discipline develops, and as it responds to particular social and cultural pressures, values and [methods] that seem paradigmatic and central can and do change" (2013, 35). Thus he suggests that core disciplinary values—not unlike those in medicine or law, for instance— necessarily undergo steady social erosion and transformation.

One exemplary transformation may be traced in the emergence and ascendance of the theory of sustainability—a buzzword that now seems to subsume all other core professional values (health, safety, and welfare) as it dominates professional discourse, design, and evaluation in many fields. Seeking an alternative to the tyranny of sustainability's growing instrumentalism, however, Christophe Girot suggests that, to be successful, genuine sustainability requires a more sensitively nuanced "both-and" plurality of values: "landscape architecture must create a new value system that is capable of combining the new economics of the city and aesthetics of a domesticated nature all in one" (2013, 16).

Girot is echoing an alternative, coexisting theory of goodness based on landscape aesthetics—one that has been beautifully articulated by many, including the current essays by Kathryn Moore and Elizabeth Meyer in this volume. In an earlier essay, Meyer provides a detailed history of the emergence of this way of thinking and explains how, in the late twentieth century, the embrace both of art and science "re-centered" landscape designers and other environmental practitioners with a dual sense of purpose:

> This chain of events—from perceiving and revealing a landscape's essential structure and character, to creating an aesthetic experience of that environment, to fostering a sense of belonging and understanding—provides a landscape architect with two important missions as an environmentalist. The more commonly accepted mission is of reflection . . . [of] existing environmental values through siting, formal gestures, and their relationship to their ecological and cultural contexts. Another mission . . . is that of projection. . . . By extension, giving significant form and meaning to ecological processes through the making of landscape experiences has laudable goals— to foster design practices that engender more mature understandings of humanity's interdependence with nature, that stir ethical as well as aesthetic debates, and that do not sacrifice landscape form in the name of environmentalism. (Meyer 2001, 243-44)

So we come back to the heart of the difficulty voiced more than twenty years ago by Rainey: "[G]iven the fact that we as individuals may espouse a wide range of differing values, is there any possibility of consensus?" (Rosenberg 1993, 38). To "find center" among such resolutely scattered positions depends on the dynamics of site-specific history, with myriad codes, practices, agendas, and actors in play. Does the very notion of "finding center" betray a naive utopianism—to imagine that it even *should* be possible to balance, integrate, or reconcile multiple values that initially appear incommensurable? Only one thing is certain: it doesn't matter any longer. Whether utopian or simply pragmatic, our human survival increasingly depends upon locating our work, our lives, and our values in that elusive space of reconciliation between the implacable poles of fundamentalism.

The Value of Professional Work to Society

What of the second, more practical, issue—evaluation? How, precisely, should environmental design professionals prove the benefits of their work toward the mandate of public "health, safety, and welfare"? In a literature review commissioned by the ASLA Task Force on Values (established in the late 1990s by the American Society of Landscape Architects), Thomas Kapper and Richard Chenoweth (2000) sought to discover an "evidentiary basis" for assertions of value added by the profession to society (149). Using codes and concepts borrowed from the field of economics ("use value" and "non-use value"), these authors analyzed research published in peer-reviewed journals in search of "objective" and "scientific" evidence of measurable value added by design. Their results were inconclusive.[8]

On the topic of health, the most reliable conclusions found in the literature reviewed were from environmental psychology, or environment and behavior fields—that "humans prefer natural settings over built environments," especially vegetated areas in urban settings (Kapper and Chenoweth 2000, 150–51). This of course carries a number of corollary ramifications in design for landscape value. With respect to safety, physical design was shown to affect crime prevention, although little direct evidence of the impact of landscape on real personal safety could be collected. Regarding welfare, Kapper and Chenoweth subdivided the topic to study the "uneasy alliance" between economic benefit and aesthetics, but found mostly anecdotal and inconclusive results (152).

In presaging the launch several years later of the Landscape Performance Series and Case Studies Investigation (lafoundation.org/research/),[9] Kapper and Chenoweth's 2000 report exhorted environmental designers to do a better job of documenting the "value added" to places through design by clearly measuring and communicating empirical results to skeptics and competitors alike (154). Recent seminars and conference panels attest that the need for better, more precise data continues to be a rallying cry for the design professions. Many design firms have embraced the strategy of "evidence-based design" (that first emerged from the specialty of hospital architecture). With regard to landscape, the "prove it" paradigm seeks to justify any and all investments, especially costs of public landscape using funds from the public coffer. Landscape services, from storm-water recharge rates to public perception and pride of place, are now increasingly being measured using both qualitative as well as quantitative measures. Which disciplines or fields are taking the greatest responsibility for the conceptual development of values-based landscape theory and practice? And how might landscape-design agents begin to synthesize these metrics connecting landscape and values in order to establish a meaningful practical and theoretical trajectory?

Professional Agency: A Values-based Agenda

The question was put to me directly by a colleague: "So where do we go from here?"

What he really wanted to know, however, was "why does this subject matter to my practice?" Let's ask it another way. Will attention paid to values bring new relevance to the field of practice? It should. Could or should a heightened awareness of values advance our research agenda in landscape architecture and other professional disciplines in environmental design? Perhaps. Should designers insist on keeping to the higher moral ground if and when it costs them work? Not likely.

Despite the complexities and contingencies that undoubtedly face a values-based professional practice, a targeted landscape agenda probably needs to be developed before any progress can occur. For a start, that agenda might include four goals: values-based landscape literacy, accountability, research methods, and professional self-regulation.

1. *Values Literacy.* Before anything else, landscape designers and other environmental practitioners must be able to *see* how landscape values operate, precisely so they can make these processes visible for and with others, especially clients and constituents. Achieving values literacy takes practice, and is best done collectively and experientially (as opposed to any authoritative narrative). Values begin at home. Students can be challenged to consider the tradeoffs between value

sets expressed in the smallest and most ordinary decisions, attitudes, practices, and habits. Teachers (at all levels) can train their students to "read" values in built landscape as well as in the practice designs they make in school. Many theories of site reading may prove vital and useful for this agenda; cultural materialism, as explored in this book, is only one of these possibilities.

2. *Values Accountability.* Any measure of accountability should include a clear understanding of the hidden costs of business as usual and the potential benefits of a values-aware practice. Myriad theories of goodness are assumed at every stage of design operation and every scale of impact, and they rule our choices—too often without our conscious awareness. Clearly naming these assumptions can help us recognize how they operate. Whose interests are primarily served in design or planning projects? A core concept in accountability is "cui bono"—who benefits? Using a "transactional analysis" (Zube 1987) as a technique in both teaching and design practice might help reveal the potential costs and benefits that accrue to landscapes, humans, and nonhuman actors and open opportunities to modify the balance sheet.

3. *Values-based Research Methods.* "What" and "how" questions of values circle back to the state of professional knowledge and its research methods. Let's recall that normative principles and practices of water conservation, stormwater collection, use of native plants, and public participation were once emergent—and not so long ago, either. Values-based design principles may be inculcated by parallel means. Objectives might include the assembly of a set of research case studies and precedents for values-specific knowledge, training for skills in value-based negotiation, and recognition/reward systems. Both the failure and the success of values-based decisions have much to teach, adding to the basis of professional knowledge as well as the power of persuasion.

4. *Professional Self-Regulation.* Professional ethics are already integrated within certification processes, and academic institutions are responsible for teaching ethical research practices. Some advocacy organizations already recognize and reinforce values-based applications in design and planning.[10] Yet values-specific problems should become more significant components in professional curricula worldwide. Values-based rubrics for

design professionals could be developed for licensing examinations; seminars on values could satisfy continuing education requirements; standing committees of professional organizations should monitor values-based research in their respective fields; and so on. And of course, in any incremental transformation toward a values-based practice, the "gatekeepers" (that is, design critics, competition review panels, boards of trustees, and funding agencies) also play a role.

Conclusion

Today, many interdependent fields in environmental design have a golden opportunity to define the terms of "goodness" for what they do—basically, to produce the yardstick by which they wish to be evaluated. To the point: if sustained discourse has any power to shape disciplinary scope and responsibility, then the conversations begun in this volume need to be continued at a larger scale, with larger groups, higher intensity, and greater clarity on mission and strategies, performance requirements, and lessons learned. Robert Riley once asserted, "[A]s landscape architects, we are agents of change. We are ultimately not just students of the world but shapers

of the world. Our maturity as an academic discipline, in distinction from a profession, lies not in journals nor doctoral programs *per se,* but in forming and asking our own questions: questions amenable to research or scholarship, to interpretation or speculation, but questions aimed at producing answers that will make us better shapers of the human habitat" (1985, iii).

This volume is aimed precisely at meeting Riley's challenge—not just for landscape architects or design professionals but also homeowners, citizens, consumers, and all agents of landscape change—to recalibrate the scale and trajectories of the questions we can and should be asking. The most important thing that all of us can do is to help make personal landscape values publicly visible, especially in the context of shared decision-making about the world's resources. If our own values remain opaque to us, then at best we can only serve as instruments of an invisible agenda. On the other hand, once conscious of the sociocultural context of our practices as scholars, teachers, activists, consumers, and citizens, we can then begin to exercise critical agency as decision makers and landscape leaders. We offer these essays, and many others cited within, in the hope of stimulating readers' capacity to sustain lively values-based discussions over the course of a semester, a landscape project, or a career.

NOTES

1. Kudos to Lake Douglas for recalling the seminal work by Christopher D. Stone (1972; 3rd ed. 2010), *Should Trees Have Standing?* Stone addresses the ontological difficulties involved in establishing legal protections for *things*—including landscapes and other members of the natural environment—whose principal purpose (conventionally perceived) is to serve the interests of those already in possession of rights and power over it. Such arguments seem "unthinkable," Stone explains, because "all [society] could see was the popular 'idealized' version of *an object it needed.* . . . [T]here will be resistance to giving the thing 'rights' until it can be seen and valued for itself; yet, it is hard to see it and value it for itself until we can bring ourselves to give it 'rights'—which is almost inevitably going to sound inconceivable to a large group" (8–9; original emphasis).

2. For example, the United States has yet to ratify any part of the United Nations Framework Convention on Climate Change, including the 1997 Kyoto Protocol, 2009 Copenhagen Accord, and 2010 Cancun Agreement, all international treaties that set limits on greenhouse gas emissions.

3. CLARB acknowledges that the term "public" is equally problematic, as any theory of goodness seems to imply "multiple publics." Contrasting legal notions (Black's Law Dictionary) of a "public at large" with the idea of a special interest group or individual, a hierarchy of interests/values results: "the public at large, a whole community, limited classes, and individuals . . . imply[ing] that the interests of a larger community may often trump those of a smaller one," as well as, presumably, a less vocal or visible one. Thus CLARB cautions us to "beware of any simple utilitarian hierarchy that equates greater numbers with greater good" (ERIN 2010, 8–9).

4. The CLARB findings rest squarely on data produced within the field of landscape architecture, thus representing how landscape architecture imagines its value, rather than its value as specifically perceived, experienced, or measured by constituents served. This does not negate the original study; rather it begs for additional work to be done.

5. Rainey's comments emerged during a 1992 panel session based in part on his responses in the article, "Most Important Questions," published in *Landscape Journal* (Fall 1992). The panel session was the centerpiece of the 1992 conference entitled *Design + Values,* hosted by the University of Virginia and the Council of Educators in Landscape Architecture.

6. It should be noted that the sample itself was already restricted by what the editors of *Landscape Architecture Magazine* tended to value during that period.

7. There is a relatively small but growing literature on teaching and learning values in landscape architectural education, ranging from social issues of race and design (Barton 2001) to new models for service learning (Angotti et al. 2011). *Landscape Journal* has hosted several theme issues with a consideration of values in education, ranging from cross-cultural learning (Hill 2005; Chang 2005) to problems in metropolitan ecology (Musacchio 2008).

8. Although values seem to be subsumed into the *ASLA Code of Professional Ethics* (adopted 1995) and *Code of Environmental Ethics* (adopted 2000), further iterations of Kapper's and Chenoweth's research are probably due (ASLA 2014).

9. Similar to the Urban Land Institute (uli.org) and other groups, the Landscape Architecture Foundation has shouldered the important role of meta-analysis for a growing number of case studies produced within a shared format with measurable outcomes. Performance benefits (for example, stormwater retention or economic revitalization) can be compared within specific project types (housing, or waterfronts). See lafoundation.org/research/case-study-investigation/.

10. The Green Building Council's LEED program (www.usgbc.org/leed) and the Sustainable Sites Initiative (www.sustainablesites.org) are both good examples of programs that guide, showcase, and reward sustainable values when exercised through responsible choices of materials and other design operations. In such systems where values-based evaluation processes succeed, undoubtedly we will see a growing canon of values-specific exemplars.

REFERENCES

Angotti, Tom, Cheryl Doble, and Paula Horrigan, eds. 2011. *Service-Learning in Design and Planning: Educating at the Boundaries.* Oakland, CA: New Village Press.

ASLA Code of Professional Ethics (adopted 1995) and *ASLA Code of Environmental Ethics* (adopted October 2000). www.asla.org/Leadershiphandbook .aspx (accessed June 30, 2014).

Barton, Craig E., ed. 2001. *Sites of Memory: Perspectives on Architecture and Race.* New York: Princeton Architectural Press.

Brown, Brenda, Robert Riley, J. D. Hunt, Leonard Mirin, Reuben Rainey, Achva Stein, and Michael Stern. 1993. "What Should We Be Asking . . . Or Was That the Wrong Question?" In *Design & Values: CELA 1992 Conference Proceedings,* ed. Elissa Rosenberg. Vol. 4:37–44. Charlottesville: University of Virginia.

Chang, Shenglin. 2005. "Seeing Landscape through Cross-Cultural Eyes: Embracing a Transcultural Lens toward Multilingual Design Approaches in the Landscape Studio." Special issue of *Landscape Journal* 24:2 (Fall): 140–56.

CLARB (Council of Landscape Architectural Registration Boards). 2009. "Mission and Goals." www.clarb.org/about/Pages/MissionandGoals.aspx (accessed July 30, 2014).

Crewe, Katherine, and Ann Forsyth. 2003. "LandsCAPES: A Typology of Approaches to Landscape Architecture." *Landscape Journal* (Spring): 37–53.

ERIN Research. 2010. *Landscape Architecture and Public Welfare (Foundation Paper by ERIN Research, Inc. for Council of Landscape Architectural Registration Boards).* Reston, VA: CLARB. www.clarb.org/Documents/ Welfare-execsummary-public-v1.pdf (accessed September 29, 2010).

Girot, Christophe. 2013. "Immanent Landscape." *Landscape Architecture's Core?* Special issue of *Harvard Design Magazine* 36 (Summer): 6–16.

Gore, Al. 1992. *Earth in the Balance: Ecology and the Human Spirit.* New York: Rodale, Inc.

———. 2006. *An Inconvenient Truth: The Planetary Emergency of Global Warming and What We Can Do about It.* New York: Rodale, Inc.

Hill, Margarita. 2005. "Teaching with Culture in Mind: Cross-Cultural Learning in Landscape Architecture Education." Special issue of *Landscape Journal* 24:2 (Fall): 117–24.

Kapper, Thomas, and Richard Chenoweth. 2000. "Landscape Architecture and Societal Values: Evidence from the Literature." *Landscape Journal* 19:2 (Fall): 149–55.

Lewis-Burke Associates. 2013. "Lewis-Burke Federal Update: Earth and Environmental Sciences" (proprietary white paper). Washington DC, June. 14 pp.

Meinig, Donald W. 1979. "The Beholding Eye: Ten Versions of the Same Scene." In *The Interpretation of Ordinary Landscapes: Geographical Essays,* ed. D. W. Meinig, 33–48. New York: Oxford University Press.

Meyer, Elizabeth K. 2001. "The Post–Earth Day Conundrum." *Environmentalism in Landscape Architecture,* ed. Michel Conan, 187–244. Washington, DC: Dumbarton Oaks.

———. 2008. "Sustaining Beauty: The Performance of Appearance." *Journal of Landscape Architecture* (Europe), vol. 3 (Spring): 6–23.

Musacchio, Laura R. 2008. "Metropolitan Landscape Ecology: Using Translational Research to Increase Sustainability, Resilience, and Regeneration." Special issue of *Landscape Journal* 27:1 (Spring): 1–8.

Nassauer, Joan Iverson, Mary V. Santelmann, and Donald Scavia, eds. 2007. *From the Corn Belt to the Gulf: Societal and Environmental Implications of Alternative Agricultural Futures.* Washington DC: Resources for the Future Press.

NCEES (National Council of Examiners for Engineers and Surveyors). 2014. "Vision, Mission and Strategic Plan." ncees.org/about-ncees/vision-mission/ (accessed July 30, 2014).

The President's Climate Action Plan. 2013. www. whitehouse.gov/sites/default/files/image/ president27sclimateactionplan.pdf (accessed May 20, 2014).

Riley, Robert B. 1985. "Foreword." *Proceedings of the 1985 CELA Conference: Prospect, Retrospect, Continuity,* ed. Brian Orland, iii. Champaign: University of Illinois at Urbana-Champaign and the Council of Educators in Landscape Architecture.

Riley, Robert B., and Brenda Brown. 1992. "Most Important Questions." *Landscape Journal* 11, no. 2 (Fall): 160–81.

Rosenberg, Elissa, ed. 1993. *Design & Values: CELA 1992 Conference Proceedings.* Vol. 4. Charlottesville: University of Virginia.

Stone, Christopher D. 2010. *Should Trees Have Standing? Law, Morality and the Environment.* 3rd ed. New York: Oxford University Press.

Thompson, Ian H. 2000. *Ecology, Community, and Delight: Sources of Values in Landscape Architecture.* New York: Routledge.

———. 2013. "Essence-less Landscape Architecture and Its Extended Family." *Landscape Architecture's Core?* Special issue of *Harvard Design Magazine* 36 (Summer): 24–35.

UNDP (United Nations Development Program) and Khalid Malik. 2014. *Sustaining Human Progress: Reducing Vulnerabilities and Building Resilience.* Human Development Report 2014 Summary. New York: UNDP. hdr.undp.org/sites/default/files/ hdr14-summary-en.pdf (accessed July 30, 2014).

Zube, Ervin H. 1987. "Perceived Land Use Patterns and Landscape Values." *Landscape Ecology* 1 (1): 37–45.

CONTRIBUTORS

JENNIFER D. W. BRITTON is assistant professor in the Landscape Design Program at Montana State University, where she teaches values-based design, theory, visual communication, and plant materials. She holds a master's degree with a certificate in historic preservation from the University of Georgia and two undergraduate degrees in environmental design and studio art from the University of California, Davis. She has previously worked for the City of Seattle as a transportation planner, a landscape designer in private firms, and an advertising art director in San Francisco. Her current research explores cultural landscape interpretation and visual communication.

KYLE D. BROWN is director of the Lyle Center for Regenerative Studies and professor of landscape architecture at California State Polytechnic University, Pomona. His current research interests examine the relationship between social justice and environmental sustainability. He holds a bachelor of landscape architecture degree from the University of Minnesota as well as a master's in landscape architecture and a doctoral degree in regional planning from the University of Massachusetts, Amherst.

M. ELEN DEMING is professor of landscape architecture at the University of Illinois, Urbana-Champaign, where she teaches landscape history, research design, and design studios. From 1993 to 2008, she taught at the State University of New York College of Environmental Science and Forestry in Syracuse. Coeditor of *Landscape Journal* from 2002 with James F. Palmer, and sole editor from 2006 to 2009, Deming also served as president of the Council of Educators in Landscape Architecture (2009–12). She and coauthor Simon Swaffield published *Landscape Architecture Research: Inquiry/Strategy/Design* (2011).

MARTIN J. HOLLAND is assistant professor in the Department of Landscape Architecture at Clemson University. His research interests include memory and memorialization in the built environment. He earned his undergraduate degree in philosophy from Dalhousie University (Halifax, Nova Scotia), a master of landscape architecture degree from the University of Virginia, and a doctorate in landscape history and theory from the University of Illinois, Urbana-Champaign. He previously worked for EDAW (now AECOM) in Atlanta and taught at the Illinois Institute of Technology, Chicago.

RAJAT KANT received his master of architecture degree from the University of Illinois, Urbana-Champaign, in 1991. He has a design practice in Lucknow, India, and teaches urban design as a visiting faculty member at Gautam Buddha Technical University, Lucknow. Kant's main research interest is the urban structure of the city and its evolution over time.

ALAN E. LONDON has been a practicing lawyer for more than forty years, most of them as a corporate partner of Reed Smith LLP. Holding bachelor's and law degrees from Yale University, he earned a master's in landscape studies from Chatham University and is currently pursuing graduate studies

in the history of art and architecture at the University of Pittsburgh. His research explores the uses of figural sculpture in twentieth-century commemorative landscapes.

TOM MELS is associate professor of human geography at Uppsala University (Gotland campus), Sweden. He has taught courses on resource geographies at Gotland University, Lund University, and Kalmar University, also in Sweden. His research focuses on landscape, environmental justice, modernity, politics, and cultures of nature conservation, especially regarding the island of Gotland. One of his current projects examines relations of planning and power in relation to wind energy production on the island.

ELIZABETH K. MEYER, Merrill D. Peterson Professor of Landscape Architecture and Edward E. Elson Professor of Architecture, is dean of the College of Architecture at the University of Virginia. She has lectured and published widely on the practice and theory of contemporary landscape design. Notable works include "The Post–Earth Day Conundrum: Translating Environmental Values into Landscape Design," in *Environmentalism in Landscape Architecture* (Dumbarton Oaks, 2001); "Sustaining Beauty: The Performance of Appearance," in the *European Journal of Landscape Architecture* (Spring 2008);

and "Slow Landscapes: A New Erotics of Sustainability," featured in *Harvard Design Magazine* (Winter 2010). Meyer's teaching and scholarly interests focus on three areas: modern landscape theory, contemporary practice of landscape criticism, and the idea of site interpretation.

DON MITCHELL is Distinguished Professor of Geography at the Maxwell School of Syracuse University and winner of a MacArthur "Genius Grant." With a focus on cultural and historical geography of urban space, landscape, and labor, Dr. Mitchell believes that "scholarship and political commitment cannot be divorced." He examines the production of landscape in relation to working people in *The Lie of the Land: Migrant Workers and the California Landscape* (1996) and *They Saved the Crops: Labor, Landscape and the Struggle over Industrial Farming in Bracero-Era California* (2012), and he coedited a collection of essays with Kenneth Olwig titled *Justice, Power and the Political Landscape* (2009). He is also the author of *The Right to the City: Social Justice and the Fight for Public Space* (2003).

KATHRYN MOORE is professor of landscape architecture in the School of Architecture, Birmingham Institute of Art and Design, Birmingham City University. Past president of the Landscape Institute and 2008 Thomas Jefferson Visiting Chair at the University of Virginia, Moore has lectured and published extensively on design quality, theory, and education. She is author of *Overlooking the Visual: Demystifying the Art of Design* (2010) as well as "The Nature Culture Divide," in *Ecological Urbanism,* Proceedings of the Ecological Urbanism Conference for 2011. She is currently advising UNESCO on the feasibility of an international landscape convention.

STEPHEN SEARS is associate professor of landscape architecture at the University of Illinois, Urbana-Champaign. He teaches studios ranging from foundation site design to advanced-level urban design and seminars about theory, practice, media, and culture. He maintains an agenda for practice that includes design for marginal urban territories, techniques in new media, and studies of the vernacular culture of the Midwest. Sears holds a bachelor of science degree in landscape architecture from Purdue University (1992) and a master's of landscape architecture in urban design, with distinction, from Harvard University's Graduate School of Design (2006). He is a fellow of the American Academy in Rome (2000) and a fellow of the Institute for Urban Design (2010).

CATHERINE SEAVITT NORDENSON is associate professor of landscape architecture at the City College of New York and principal of Catherine Seavitt Studio. She coauthored the book *On the Water: Palisade Bay* (2010), an infrastructural and ecological climate-adaptation proposal for New York Harbor in response to the effects of sea-level rise and storm surge. Seavitt's current research examines landscape restoration practices given the dynamics of climate change; she is currently investigating the history, processes, and ethics of the de-domestication of large herbivores for grassland restoration and land management.

AMITA SINHA is professor of landscape architecture at the University of Illinois, Urbana-Champaign, where she has taught since 1989. She has studied the cultural landscape of Lucknow, India, for the last twenty-five years, beginning with her doctoral research on social aspects of low-income housing. She has authored numerous articles on historic gardens and contemporary parks, historic streetscapes, riverfront revitalization, and neighborhood spaces of the city. Sinha is also the author of *Landscapes in India: Forms and Meanings* (2006), and editor of *Landscape Perception* (1995) and *Delhi's Natural Heritage* (2009). She was a senior Fulbright research scholar in 2009.

INDEX

Note: Italicized page numbers indicate illustrations and their captions.